ETERNITY SOUP

ALSO BY GREG CRITSER

*Generation Rx: How Prescription Drugs Are Altering
Americans' Lives, Minds, and Bodies*

Fat Land: How Americans Became the Fattest People in the World

ETERNITY SOUP

Inside the Quest to End Aging

GREG CRITSER

Harmony Books

NEW YORK

Harmony Books is a registered trademark
and the Harmony Books colophon
is a trademark of Random House, Inc.

Library of Congress Cataloging-in-Publication Data

Critser, Greg.
Eternity soup / Greg Critser
p. cm.
Includes bibliographical references and index.
1. Longevity.   2. Aging—Prevention.   I. Title.
RA776.75.C75 2010
613.2—dc22
2009030670

ISBN 978–0–307–40790–0

Printed in the United States of America

Design by Elizabeth Rendfleisch
Illustrations by Andrew Barthelmes

1 3 5 7 9 10 8 6 4 2

First Edition

In memory of Paul Clyde Critser, beloved father

CONTENTS

"So why do we complain about nature? She has acted kindly: Life is long if you know how to use it."

—SENECA

"Old age ain't no place for sissies."

—BETTE DAVIS

INTRODUCTION

We were lunching in God's waiting room, which is what Bob Hope liked to call Palm Springs, when the subject of aging came up.

It was a good table for the subject, because if you were looking for two rational examples of healthy aging, you probably couldn't do better than my mother, Betty, seventy-six, and her husband, Jerry, seventy-five. They've got it down: a simple, sensible diet with an occasional splurge, routine daily exercise, some yoga, lots of socializing, and lots of mental stimulation via travel, books, hobbies, and adult education. They're the geriatric dream team—sensible, moderate, willing to try new things, and proactive when it comes to seeing their kids and grandkids. They're not going to sit and ruminate over *your* not coming to see them. They'll come to see *you.*

They are also huge and fairly discriminating consumers of health and medical information, and so it was with some relish that, one day over lunch not long ago, I told them about some new research I'd been reading on aging and health. "You see," I said, dropping into PowerPoint mode, "the real issue isn't life span, it's *health* span. And what do we know about that? Well, we know that healthy aging is just that—and that what medicine should aim for is lengthening the *health* span, not the life span."

They looked at me a little blankly and picked at their salads.

I went on. "And so, that's led to kind of a new Holy Grail in aging research—the idea of rectangularization of morbidity."

More blank stares. Forks now on plate.

". . . 'cause, see, if you look at aging demographics, you see that the survival curve goes like this . . ."

I traced a downward-sloping line on the tablecloth with my finger.

". . . with increasing sickness and disease as people get older. What we really want to do is make it square off, like this, at, say, age eighty-five."

I traced a ninety-degree angle, downward.

"What does that mean?" my mom asked.

"Oh, simple," I said. "It means that the goal of medical and aging research should be to get you to, say, eighty-two or eighty-five in good health, and then you drop off fast, say, in three to six months, without extended illness."

There was a silence, and I knew I was in trouble.

"Oh," my mother said, "I don't go along with that at *all*. . . ."

"Uh-uh," said Jerry. "That's not how we're seeing it at all."

"The way I see it, I'm going to live to one hundred and maybe more," my mother said. "Why not? And a healthy hundred at that!" She picked up her fork and—nicely—stabbed an artichoke heart to death.

I stifled myself from pointing out that their gene pools did not augur well for such optimism, but, at this point, they owned the conversation. I wasn't taking into consideration something they both seemed to know a lot about: antiaging medicine. They both had a "longevity doctor." Their treatments included prescription "compounded" hormones, testosterone for him and progesterone cream for her. There were all kinds of things you could do to slow down the aging process.

"But, Ma," I said, "you know that there's no evidence that that stuff does any more or less for you than taking prescription hormones!"

"Ah! But there is! The difference is the cream is tailor-made for me! I get a blood test every month to make sure. And Gregory—I know you wrote a book about all this—but I am telling you it works! My skin hasn't been like this for twenty years! And my energy level too. And I'm not even doing my yoga as much, so I know it's the cream. It's definitely making me feel younger. And frankly, Gregory, I don't even think of aging the same way anymore. Aging, like that I mean, when you really think about it, it's so unnatural!"

Driving home that day, I thought a lot about that last statement, the notion that aging was somehow unnatural. It was the kind of dimensional shift in perception that leads you to think that, once again, just like that "crazy Internet thing," modern life has passed you by. Either that, or that everyone around you has gone totally off the grid.

After all, everyone knows that antiaging medicine is bogus, right? Certainly all the right-thinking folk think so. Even the somnolent FDA recently raided a bunch of pharmacies for selling antiaging compounds. The Gerontological Society of America—let's call it Big G—sponsors seminars on antiaging quackery. A group of Big G researchers has even taken to issuing consensus edicts against antiaging medicine, often noting its more obvious frauds while impaling the whole notion as quackery. Some of these attacks are strangely personal, singling out bona fide researchers who differed from the establishment line as "rogues" and impaling others for being "celebrities."

But something has bothered others in the fields of aging and medicine about this consensus. When you looked a little harder, you saw that some of the bigger names had refused to sign on. It was too soon to take such a censoring position, they said. Younger scientists viewed it as a generational affront. As a graduate student named Adam Spong, a cell biology researcher at Southern Illinois University Medical School, told me, "Their whole thing is just so, so . . . postmodern in that 1970s way. I mean, you cannot talk about aging as a disease, because that would be ageist. You cannot talk about something objectively factual—that it is better, physiologically, to be young than be old—without supposedly causing stigma or victimizing the elderly. To me it all stinks of some weird, self-defeating orthodoxy, call it political correctness, if you want."

In fact, the more one looks at antiaging, or longevity, science and medicine, the more one sees that there is no real consensus about it at all, and that myriad enterprises, both public and private, are forging onward into a brave new pro-longevist world. Consider, for example, that:

•   Stanford University professor Shripad Tuljapurkar, an expert in population mathematics, projects that antiaging advances in developed countries could increase life expectancy *by a year for each year* of the decades 2010 to 2030.

- The National Institute on Aging, the country's federally funded arbiter of gerontology, is now underwriting a long-term, wide-ranging program to test life-span-extending compounds on mice at three leading research institutions. Although designed to debunk commercial claims, the early results have already identified three compounds that extend maximum life span in mice—a huge surprise to the dedicated skeptics running the trials.

- Another NIA study, the first on humans, showed that thickening of the carotid artery, a key sign of aging and a risk factor for stroke, was dramatically reduced in practitioners of caloric restriction.

- When the noted RAND Corporation asked the nation's leading gerontologists, cardiologists, and geriatricians about the possibility of a major life-span-extending compound, many were surprisingly bullish, with a number predicting a 50 percent chance of such a compound in the next ten years.

- The cofounder of PayPal, Peter Thiel, one of the financial world's shrewdest investors, has put $3.5 million into a project to do nothing less than "end aging."

- The pharmaceuticals giant Glaxo recently paid $750 million for the development rights to resveratrol, a compound that seems, among other things, to slow the underlying aging process in some lab animals

- The head of Rice University's prestigious environmental engineering department, considered one of the best in the world, is now leading an experiment that will use soil bacteria to target and destroy human arterial plaque, essentially creating a way, in the words of one researcher, to "end arterial aging as we know it."

- The NIH is funding trials in pig-to-primate organ transplantation, hoping to push the science closer to usefulness before current demographic trends worsen and organ demand outstrips supply even more dramatically than at present.

- Aging Americans are no longer waiting for breakthrough science or Big Pharma. A soaring number of prescriptions going out the pharmacy door these days is for a hormone-based product, most of them used for antiaging purposes. Instructively, it is almost always paid for in cash.

I come to this remarkable and fast-moving world with my own biases and prejudices. Like most people, I *want* antiaging medicine to work, especially since men in my lineage don't tend to live very long. There are loved ones all around me whom I'd like to enjoy a lot longer. And, surprise: I think the bodily and mental degradation that comes with advanced years is cruel, capricious, and unrelenting, and the attempts to maintain one's dignity in the face of it are too often clownish, pitiful and, ultimately, ineffective. A recent concussion—which resulted in a form of accelerated brain aging and for which I sought treatment in antiaging medicine after conventional medicine failed—drove that point home in a personal way. A jock would say that I've got skin in the game.

The more I looked, the more it seemed to me that, over the next fifty years, science, medicine, and technology will transform aging—and the way we think about it. This change will come in ways small and large, its evolution gradual but inexorable, fueled by both self-experimentation and publicly funded trials. Much of it will take place in terra incognita; aging may become easier and more comfortable, but more expensive and even a little riskier, too.

So where does antiaging, or longevity, science come from, what does it tell us, how does it connect (or not) with traditional medicine and medical science, and how will we think of our own bodies as a result? Those are the questions this book will try to answer. To start, we must first go to Tucson, Arizona.

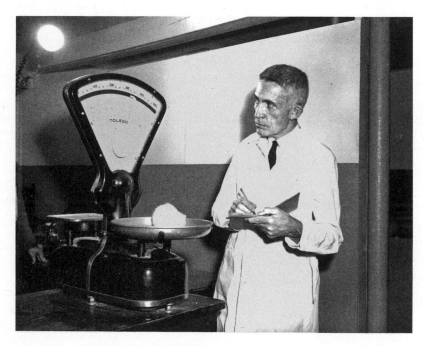

*Clive McCay, the father of modern longevity science,
and one of his subjects, circa 1930.*

BOOK ONE

Calorics

We must resist old age, Scipio and Laelius; we must compensate for its deficiencies by careful planning. We must take up defenses against old age as against a disease, taking due thought for good health, following a program of moderate exercise, eating and drinking enough to rebuild our bodies, but not to overload them. —CICERO

"Rubbish!"

Silence.

"Rubbish, I say!"

Silence.

The speaker, Josh Mitteldorf, having quietly taken in the one-word critique issuing from somewhere in the hotel conference room, commenced again. He was trying to convince a gathering of the Caloric Restriction Society, gathered for their second international conference in Tucson, Arizona, that death wasn't really such a bad thing, that nature evolved the genetic program for apoptosis—programmed cell death—for a reason: to keep the planet from getting overcrowded. "I kept asking myself," said Mitteldorf, a slim, tousle-haired environmental scientist, yoga teacher, and philosopher, "why would evolution create an organism with a program to kill cells? And a lot of the answer came from looking around at the world today—at obesity, overpopulation, pollution, consumerism—and that all led me to this quest to understand why nature made us to die." He had recently published a paper in *Science* about it, and . . .

"*Rubbish!*"

This time there was an audible ripple of protest through the crowd, directed toward the source of this indecorous behavior, a tall, skinny man wearing a T-shirt with the words END AGING NOW! His name was Michael Rae, only twenty-nine and who was, along with his equally young girlfriend, April Smith, quickly emerging as the new poster children for the CR Society, an organization long saddled with the dour image of some of its older practitioners. Rae, who had already charmed everyone by openly flossing his teeth during an earlier presentation—the better to reduce his inflammatory burden— was having none of Mitteldorf's ideas, and instead of waiting for the formal question-and-answer period, bounded up to the podium, cranked the microphone his way, and began to recite data to the contrary. "Can someone let me use their computer—I want to look up a reference?" he said. He then went on, with a barrage of . . . data. "So how can you possibly make this argument?" Rae said, winding up, to an apparently unperturbed Mitteldorf. (There had been a yoga class that morning.) "How can you *say* that?"

I had come to Tucson at the invitation of Lisa Walford, a

former president of the Caloric Restriction (CR) Society and
the daughter of one of the organization's founders, the late
Roy Walford. Tucson held some special significance both for
the society and for Walford. About an hour's ride out into the
desert stood the site of one of the first, albeit unintentional,
sites of human caloric restriction—Biosphere 2. The biosphere was an attempt to live for two years in a fully contained,
self-supporting environment. It ended in a rash of controversy,
partly because the dome failed to produce enough food or air
for its eight inhabitants. Roy Walford, one of them, had never
been the same after, and many quietly suspected that his recent death, at age seventy-nine of Lou Gehrig's disease, to be
a direct result of the biosphere experience. The scientific program of the CR conference included a number of highly placed
scientists who had been friends with Walford, here partly to
respect that, partly because, when you think about it, CR
people are the only lab animals that can talk.

"You can learn all kinds of things from them," said Edward
Masoro of the University of Texas; he had been studying CR
in rats for decades, "but there is nothing like this conference."
"These are people with a huge commitment—think about it,
they are willfully suffering, in my mind, for a payoff we don't
really know will work in humans," Steve Spindler, a cell biologist from UC Riverside, told me later. "Usually they are
eating thirty to forty percent less than they normally would—
every day. These are not crazy cult members either. Yes, they
get passionate, but they are the most scientifically well-versed
people I have ever met. They are amazing."

Luigi Fontana, a professor of medicine at Washington University School of Medicine, chimed in over a bland meal of

couscous and steamed vegetables later: "These are wonderful people—beautiful people!" He smiled and went on to outline his plans to draw their blood and study their arteries and brains. "I mean, where—where!—can a medical scientist ever hope to have anything better!"

And where, I thought, can a journalist hope to sight more fish in the same barrel? For it is true that, among CR people, there are plenty of odd characters: the vitamin marketer who eats so much beta-carotene that his palms look slightly orange; the mathematics professor who tries to get so many of his limited calories from raw vegetables that a meal with him takes two hours and a dozen trips to the salad bar; the expatriate electronics executive who normally lives in Japan and who prefers to eat his dinner off a pharmacist's scale, gram by gram; the perfectly healthy advertising executive who measures blood sugar before and after every meal; his wife, who believes that eating after two in the afternoon is "something we should just get over." There are plenty of odd events as well: the six a.m. meditation breakfast, in which one meditates on a repast of five blueberries and three potato chips; the sometimes mind-numbing scientific panels on coenzyme Q10, l-carnitine, PBN, and, of course, protein versus carbohydrate consumption. All of this had led, in the past few years, to a rash of magazine and newspaper articles, inevitably describing the strange aspects of the society and, not surprisingly, using it to vent any variant of modern identity politics, from feminism (there were almost no female members) to fatty-ism (it was a "pro-anorexia front") to foodie-ism. "I don't know what was worse, the accusation that we were somehow *for* anorexia or that we didn't know anything about cooking," April Smith said, a little sarcastically. "I mean, I take some pride in my halibut!"

Yet the more time I spent with CR people, the more I came to see them as utterly logical (perhaps supralogical, given the consumption-oriented culture we all live in today). For mainly a series of unremarkable reasons—usually set in motion by a health crisis and an early realization that they are mortal—they each have evolved a kind of longevity phenotype—a set of traits—of their own. In a sense, they represent one version of what people might be like in a world of extended life spans and ratcheted-down consumption. They are, literally, cool: Shake a hand and you'll notice. They can be a little grumpy, too, but that is just those who are relatively new to it. The rest, as Lisa Walford, who grew up in the household of the world's most famous calorie restrictor and is now one of LA's top yoga teachers, explained, "just have a kind of damped-down affect. It is not depressed. It is just . . . different." She paused for a moment. "You'll want to compare it to the mouse, of course, but what I find is people who are just not as revved up as we expect people to be these days."

They are also not very sexual, a subject to which we will return later.

All of which invites the question: Where did we get CR in the first place?

Cornaro's Way

Not long ago, Steven N. Austad, one of the nation's foremost experts on the evolutionary biology of aging, was rummaging around in the dusty archives of Clive McCay, the Cornell University scholar credited with discovering, in 1935, the

LUIGI CORNARO

Dal ritratto dipinto da Tiziano

Alvise "Luigi" Cornaro, sixteenth-century author of La Vita Sobria, *which influenced modern antiaging science.*

well-known antiaging effects of caloric restriction. Among the yellowing photos, old laboratory notes, and studies on nutrition was McCay's journal. On several pages McCay had jotted down tips for healthful eating, a few aphorisms, and a piece of recommended reading: a book called *La Vita Sobria: How to Live a Long Life*, by a sixteenth-century Italian, Alvise "Luigi" Cornaro.

Right away Austad, whom I'd interviewed and who knew of my own interests, sent me a copy. "Here it is," Steve wrote. "I see he's reading your friend here." In his usual low-key way, Austad had made an important find. He had linked the most important science about life extension of the twentieth century to its distant philosophical origins in the sixteenth.

Cornaro, a Renaissance humanist and businessman little known outside of academic circles, had been on my mind for nearly a decade. I first came across his treatise while researching a book about obesity and was struck by the modern arc of his health narrative. Here was the classic case of a man falling ill because of his wayward behavior, then regaining his health through renewed attention to diet and vigorous living, and then — the modern part — prescribing his regimen to every friend who would listen, first with incessant chatter at the dinner table, then via a self-help book. Cornaro was our kind of guy: the original billionaire philanthropist turned health nut.

Or was there more? As I slowly picked through translated Italian documents and visited his home in Padua and his farmlands just outside Venice, something else began to emerge. The more I looked, the more I saw that Cornaro's circumstances, in terms of bodily aging, were very much like our own. True, they included madrigals and *commedia* and busty wenches and ermine stoles, the Renaissance version of MTV and Internet porn, but if you parsed his physiological world through modern scientific lenses, you came away with a prototype for contemporary aging—and for contemporary antiging.

Born in Venice in 1484, Cornaro spent much of his early life apprenticed to a wealthy uncle in Padua, who put him through school and college and, perhaps more important, taught him how to work the ropes of the Venetian bureaucracy for financial succor. At the time, Venice and her empire had entered a period of almost constant warfare with the Holy Roman Empire; her preeminence in international trade suffered huge blows. Cornaro—young, ambitious, a fast study—took this all in. The future, he came to believe, lay not in overseas conquest, but in radical agricultural investment at home. Inheriting some fetid marshland on the mainland from his uncle, young Luigi drained it and planted rice, wheat, and vegetables. He must have done something right. The swamps burst with plenty, and he made a fortune—a fortune that launched him into a round of partying that didn't end until he turned thirty-five. He came to his physician debauched and depressed; he was told he'd likely not make it to his next birthday.

His symptoms were telling. Cornaro had "a continuous unquenchable thirst." He could not tolerate sweet fruits. Or pastry. He was also gouty—probably from too much fatty meat. Today, we would likely say that he had type 2 diabetes. (His case pre-dates that of J. S. Bach, long thought, albeit uncertainly, to be the first *named* type 2 sufferer.) Cornaro also complained of seasonal fevers, likely from low-grade malaria contracted while surveying his swamplands. In the Galenic parlance of his day, Cornaro considered himself "choleric"— eternally hot and bothered.

His physician, perhaps having parsed some of the period's reissued classical medicine texts, gave him an unusual prescription. He had to learn to eat what agreed with him, but—most important—*to stop eating before he was satisfied.* Much of the advice was couched in moral terms, with great emphasis put on the notions of temperance and sobriety, the divine counters to gluttony and the way of all evil. (This was before postmodernism, when you could still speak about such matters.) But as Cornaro himself discerned it, the advice was the essence of reason as well: "I accustomed myself to the habit of never fully satisfying my appetite, either with eating or drinking— always leaving the table well able to take more. In this I acted according to the old proverb: *'Not to satiate oneself with food is the science of health.'* "

Within a few days he felt better, and he began to experiment. Fruit did not agree with him, and despite the prestige accorded to melon and cherries and apples and peaches on the rich man's Renaissance table, he simply stopped eating them. The same with pastries. Meats like goat and mutton, in small amounts, he found agreeable. *Pane padovano*, the coarse, whole-grain country bread of the Paduan countryside, he found a delight. But Cornaro's mainstay was his "dear" *panatella*, or, as it is known today, *panado*—a brothy soup, usually derived from capon and cooked with small bits of Paduan bread and, sometimes, an egg. With it he drank two glasses of wine—"the milk of the aged," he called it—a day, bringing his total daily intake to somewhere between 1,500 and 1,700 calories. Thus did a man who never heard of blood sugar or insulin, let alone diabetes, put together a relatively low-sugar, high-protein, calorie-light diet. Cornaro's Way might suit many modern type 2 diabetics just fine.

The results of his regimen were striking. "In less than a year, I was entirely freed from all the ills which had been so deeply rooted in my system as to have become almost incurable." A man transformed, Cornaro could now mount his horse "and other things" unaided. He took up anew his vast designs for transforming Venice (he devised an influential plan to block off the lagoon from the ocean and build an island theater across from St. Mark's, which later critics considered heresy), and he began writing influential tracts on architecture, water management, and farming. His own garden became the scene of one of the liveliest and most influential salons of the northern Renaissance. The great language master (and literary inamorata of Lucretia Borgia) Cardinal Bembo partied, apparently soberly, with him. Jacobo Tintoretto came and painted Cornaro's portrait (which now hangs in the Pitti Palace). Giorgio Vasari interviewed Cornaro. Even the young Andreas Palladio, later the preeminent architect of the northern Renaissance, absorbed Cornaro's enthusiasm for classical architecture. The old man's Casa and Loggia Cornaro—filled with racy murals and tableaus right out of Nero's Golden Room—anticipated Palladio's revival of classical Roman architecture by more than a dozen years.

Moreover, Luigi Cornaro was aging with grace. "I am healthy, cheerful and contented," he wrote in 1558. "My sleep is sweet and peaceful and moreover, all my faculties are in a condition as perfect as they ever were. My mind is more than

ever keen and clear, my judgment sound, my memory tena-
cious, my heart full of life." He was happy, enjoying his days
spent with his eleven grandchildren, who lived in his home.
With them he often sang aloud. "My voice—that which is
wont to be the first thing in man to fail—is so strong and so-
norous that, in consequence, I am obliged to sing aloud my
morning and evening prayers, which I had been accustomed to
say in a low and hushed tone. Oh, how glorious will have been
this life of mine! O divine temperance!" When he wrote those
words he was seventy-six—more than twice the age he was
when given his death sentence forty years before.

In terms of life extension and aging, what exactly had
Cornaro done? Clearly he had stumbled onto something. Today we would call it calorie restriction, or, simply, CR, which
we know works in extending the life span of mice, rats, fruit
flies, earthworms, yeast, and, many hope, humans. But how
does it work? Of all the proposed mechanisms now in vogue,
Cornaro's Way goes to the heart of the two most vibrant
theories. One holds that CR works by constantly, but mildly,
stressing the body with faminelike cues, thereby pushing cells
to stop putting limited energy into reproduction and growth
and instead invest that energy in maintenance and repair, usually repair of damage from routine metabolism; you can think
of this as the "super-repair" pathway. The other main theory
holds that CR works by ratcheting down the amount of insulin and sugar that runs around in the bloodstream, thereby
preventing diabetes—itself a form of accelerated aging—and
cardiovascular diseases and a number of age-associated cancers. Think of this as the "sugar signal" pathway. These routes
are far from mutually exclusive, with one often affecting the

other. Cornaro himself had his own idea about why it worked; in Galenic terms, "eating little" stanched the slow but inevitable loss of "radical moisture," something that every person possessed in but a fixed portion.

Yet, ultimately, it did not matter why CR worked. Cornaro, surrounded by beautiful singing grandchildren, a garden full of trendy intellectuals drinking divine milk, had tripped headfirst down a physiological pathway only now being deliberately parsed by the leading minds of American science. O divine temperance!

And so: in 1558, he published *La Vita Sobria*, with new editions following in 1562 and 1564. Contemporaries wrote paeans to him and to it; his friend and language scholar Sperone Speroni even wrote an Erasmus-like gloss on it, "In Praise of Gluttony." (If it all took place now, Cornaro's home would probably be featured in something like *Simply Rinasciemento!* magazine.) And then, in 1566, at the age of eighty-three, Alvise Cornaro died; according to one friend, there was no pain, no regret, no tears. The old man, sitting in his little bed, placidly drifted off, "as serene as a beautiful sunset on an unclouded day."

In short, his survival curve had been rectangularized.

But what of Cornaro's Way? Its author had departed, but the book—it had stuck. *La Vita Sobria* quickly traveled the corridors of power in Europe—to France, Holland, Spain, Germany, and England. And then across the Atlantic. In 1793, the Reverend Parson Weems, arguably George Washington's most important hagiographer (he literally invented the myth of the cherry tree) and an enterprising upstart publisher, packaged a translation of *La Vita* with an essay on health from

Benjamin Franklin; it became a postcolonial must-read when he convinced the new president to blurb the book, now retitled *The Immortal Mentor.* So did Cornaro's Way enter the health literature of the New World.

In 1903, Thomas Edison blurbed a tony, gold-edged reissue.

In 1920, Henry Ford, the ultimate American millionaire health nut, gave it to friends as a Christmas present.

Growth is death: The farm boy, the brook trout, and the white rat

Just outside Burlington, Vermont, flows the Farmington River, which sends its burbling brooks coursing through the bucolic New England countryside, and with them, the aptly named brook trout, or *Salvelinus fontinalis.* They are a beautiful fish, legendary for a good fight and, after a cold morning spent in hip waders, an even better fry. They were also, scientifically speaking, America's first, albeit involuntary, practitioners of caloric restriction. Today, you can get pretty close to these longevity pioneers: In noisy 2008, you can still stand in the exact spot where modern antiaging science was born, the place Cornaro met the Connecticut Yankee. That spot, just off Route 4, is the Burlington Trout Hatchery. Wind your way through the circular pools and on into the wooden hatchery building, and you will find a series of troughs, meant for processing young trout, or fingerlings, before release. It was here, sometime in August 1926, that the biochemist Clive McCay and his assistant, F. C. Bing, reveled in the wonders of nature, science, trout farming—and life extension.

McCay was a young man, having just earned his PhD from UC Berkeley, but already he was a piece of work. An Indiana farm boy, orphaned as a teenager, McCay came to college with calloused hands, having worked summers as an itinerant wheat shearer. Long, tall, craggy faced, and handsome in a Nordic way, he was also the classic fast study. As a boy he inhaled the natural sciences, and one day, having gotten his hands on a government bulletin about nutrition, "I learned about calories," he recalled later. "My sister says there was never a calm meal thereafter because I always sat down and counted the calories in the potatoes and the bread." McCay spoke quickly and was always moving—"Indian-like," as one friend put it—hiking ridges and hollows, botanizing strange and exotic stands of grass and weeds, and fording dark, mysterious streams. And there were McCay's omnipresent companions—terriers! beagles! retrievers!—which seemed to follow him, playfully nipping at his heels, partly in pure canine joy, partly in stoic mammalian reverence for the master, who seemed always to know the way home.

He could well have become a veterinarian, were it not for the fact that, in the mid-1920s, one of the hottest subjects in the sciences was nutrition. As a postdoc, Clive McCay took a fellowship at Yale to study animal nutrition with the great L. B. Mendel, credited with isolating vitamins A and B, among other things. There, in 1926, McCay met F. C. Bing, a young Yalie fascinated with, as he liked to call it, "the vitamin sciences." The vitamin sciences were popular because, after years of frustration, there was a tangible payoff to it all. That same year, the medical researchers Ming and Minot discovered that you could cure pernicious anemia, a seemingly intractable problem, by having a patient eat large amounts of liver. The result of the discovery was an explosion in nutrition research—and a huge increase in demand for raw liver, traditionally part of the feed for hatchery trout. McCay's job as a postdoc was to find alternate, cheaper feed mixtures for Vermont's nascent attempt to restock its barren streams with trout. Liver was simply too dear.

Right away, as Bing later recounted, "the principal observation we made was that brook trout, in addition to vitamins, require a substance which is present in raw liver and to some small extent in dried skim milk." McCay, attuned to the need for a vitamin-era name, dubbed it Factor H. But how much was enough? He experimented with various mixtures: 10 percent liver, he found, would keep the little trout alive, but 15 percent was needed for them to reach their maximum size. There was another thing. Reporting his results to McCay, Bing one day noted an odd pattern of death among the fingerlings. "Most of the fish in the group receiving only 10 percent protein, an amount that prevented them from growing, were still alive. They seemed to be in fairly good condition. On the other hand, most of the fish that received more protein and had grown steadily all summer, had died off." McCay was intrigued. Growth . . . and death.

"What do you think caused this?" he asked Bing.

Sitting on the ground outside the hatchery, the yellow-breasted chats chirping away, the pair chewed over their latest book learning: Rats stunted in one famous experiment appeared notably youthful; in their own mentor Mendel's lab, stunted rats on deficient diets for a long time could resume almost normal growth upon commencement of regular feeding; in a competitor's lab, slowly growing rats on a "poor diet" lived longer and healthier than well-fed rats on "state-of-the art" diets. Everything the young men knew about life span and nutrition seemed in flux. They fell to philosophizing and speculation. Hadn't one German philosopher compared a newborn's metabolism and growth with that of a wound-up spring? "This force is continuously used up as the spring unwinds and when the spring at last runs down, death occurs." The pair came back to the fingerlings unwinding in the hatchery. "They behave," McCay and Bing later wrote in the *Journal of Nutrition,* "as if there is something in them which is gradually consumed during growth so that if the animals are kept from growing, they live longer." Growth—it somehow hastened death. In McCay, the agile young faun of Jazz Age vitamin science, the observation prompted a radical thesis: Retarded growth and development were fundamentally tied to the life-extending effects of dietary restriction.

That afternoon, McCay told Bing that he knew what he was going to do with his career. He "was going to work on longevity," and he was going to do it with . . . rats.

Rats were something that Cornell University had, and McCay, taking a post there in 1927, immediately fell to playing with their diets. The dominant lab animal of the time— mice would not topple them for another thirty years—was the white rat. It was a reliable animal model, fecund, predictable, and pedigreed; a scientist could know its lineage. But as McCay began to peruse twenty years of background data about *rattus,* he discovered an amazing fact: *almost no one knew how long they lived in the lab, let alone outdoors.* After any experiment, leftover animals were simply killed, or "disposed." About this, McCay complained and complained, until he was shut down with a challenge by Mendel: *"You're young—you* do the life span studies!" Commencing in the early 1930s, McCay did just that.

Looking at old data first, McCay calculated the average life span of the lab rat as somewhere between five hundred and six hundred days. He then designed a series of experiments to see how dietary restriction would alter growth, maturity, and life span. In his most important experiment, he took 106 weaned rats and divided them into three groups. Group I was fed all the food they desired. Group II was calorically restricted and forced to mature slowly, with feed that contained all essential vitamins and minerals. Group III was allowed to grow normally for two weeks after weaning, then forced to develop slowly with the restricted diet. By splitting the groups so, McCay hoped to figure out one of the big concerns of Great Depression nutritionists: Could growth, interrupted by lowered caloric intake, be refueled later in life? Would a slow-growing mammal—rat, cow, pig, perhaps even a human—ever reach the same size as one that matured quickly?

Although the answers to McCay's questions were somewhat predictable—slow growers never reached the size of fast growers—the health and longevity data were counterintuitive. The restricted animals lived longer—lots longer—and they had a reduced rate of disease. In reporting this, McCay became a crafty, modern communicator. Even before his scholarly journal articles on the experiment appeared, he was spinning it in the science monthlies. Writing in the September 1934 issue of *Scientific Monthly*, McCay noted: "Earlier experiments in our laboratory and [at] Columbia have shown that the mean life of a male rat is between five and six hundred days: if the average man lives fifty to sixty years, about ten days in the life of a male rat equals a year in the life of a man. In the present experiment the mean length of life was 509 days for the normally-growing male rats, equivalent to a mean age of 51 years for man. The rats of the two retarded-growth groups have exceeded mean ages of 780 and 870 days, respectively, equivalent to 80 to 90 years for man." Then, almost as an afterthought, he added: "These data indicate that the potential life span of an animal is unknown and may be far greater than we anticipated." *The potential life span of an animal is unknown and may be far greater than expected.* Again and again for nearly a decade, McCay returned to these words. Maximum life span—in human terms always thought to be somewhere around the Bible's 120 years—was not fixed. So what *was* the maximum life span? Why, by 1935 there were rats in his lab that had lived more than thirteen hundred days, and everyone knew what that meant in human years . . .

The insight seemed to transform McCay. The man once known for his nature boy profile began wearing a red bow tie, and referring to himself, as was quietly fashionable at the time, as "a bit of a Bolshevik." He was constantly speaking at this scientific conference or that, his tuft of unruly hair and handsomely rough visage looming over his peers, who enjoyed the upstart's gentle laugh and out-of-the-ordinary observations. His "growth versus longevity" thesis also dovetailed nicely with the era's prevailing theory of aging, also known as the "rate of living" thesis. To promote his line of thinking—this in the dead of the Great Depression—McCay gave interviews to popular magazines and took up the new media of the radio. His weekly show went into all kinds of detail about diet, health, and aging. He talked about how studies of nineteenth-century British prisoners showed that those who had been given the more meager rations tended to live the longest; about obesity, fat, and sugar and their effect on health. He began quoting the wisdom of the ancients and near ancient, bringing to his Cornell-area audience the words of people like the mystical thirteenth-century monk Roger Bacon, an early longevity advocate, and, of course, Luigi Cornaro.

But the more McCay talked, the more he realized his true interests lay not in animals, but in humans. (He liked to quote Bacon's injunction that "the cure of diseases requires temporary medicines but longevity is to be procured by diet.") Life span was not fixed by some law of nature, and "therefore," he said in one radio show, "a bright future belongs to the man who directs his efforts to stretching life span. We have learned to keep most of our children from dying but we have not made much progress toward giving men and women a healthier middle age. We do not wish to prolong the suffering that goes with feeble old age; we want to extend the prime of life when most of us live and enjoy living. I believe it can be done if we give this problem a sufficient amount of thought."

In the Hollywood script version of McCay's later years, he'd have likely pursued this semi-immortalist dream, slowly working his way up the mammalian chain, testing his ideas about dietary restriction on dogs and pigs and sheep and monkeys, parsing its effects and refining his theories until one day, in the great scientific breakthrough that changes the world and leaves its discoverer revered (and comfortable in tweed), he triumphs. But life rarely unfolds like a Hollywood script (and even when it does, it is more likely Antonioni than Capra). So it was for Clive McCay, who spent much of World War II studying not life span extension but the nutritional requirements for American overseas troops. It was a fruitful endeavor in itself, and when he returned, he had a list of research questions: Is drinking coffee and soda pop good for you? What kind of nutrition builds the best bone? Should drinking water be treated with fluoride? These would keep him ever busy in the lab and, more and more, in the test kitchen.

McCay's obsession in later life was building the perfect loaf of bread, an endeavor that succeeded remarkably well. By the mid-1960s, when he died, you could see that success inside the nation's small but proliferating congeries of health food stores. There, perhaps offered by a vivacious grandmother in a pink smock and sensible shoes, was a loaf of something called Cornell Bread. It was a tasty, if dense, loaf, full of whey and soy and wheat germ and, of course, powdered milk. If you wanted, you could easily make it yourself, because the recipe was inside a little book, usually on the rack next to something by Gypsy Boots or Jack LaLanne or Euell Gibbons.

The book was by a man named Clive McCay and his wife, Jeanette. If you opened it, you would encounter a strange thing—strange, at least, for a cookbook. For there, right on the fourth page, popped a big photo—of a congenial-looking, craggy-faced man in a lab coat, merrily tending two furry white animals known to most people as . . . rats!

The leisure of the theory class

One of the consequences of McCay's abandoned dietary restriction trials was to leave behind a deeply incomplete understanding of how CR worked to prolong life. McCay's idea—that CR involved something having to do with a mysterious Factor H that burned up during growth but was preserved by slowing down growth—served to attach the theory at once to the concrete *and* the mystic. If McCay had known what science would eventually come to know about B_{12}, the element in raw liver and dried milk that was essential for growth, well, then, he would have changed his theory; it was not just some nutrient that caused prolongation of life span, or everybody getting adequate vitamins would become centenarians. On the other hand, if CR's effect was somehow tied to the slowing down of growth and development, well, that process was so fundamentally governed as to be unfathomable; he didn't have any way to locate genetic mechanisms, as we do today. And so CR, tied to McCay's ideas about its workings, was stuck. Growth— somehow it led to death. Such was the notion that professional gerontology inherited in its most formative years in America.

Gerontology—the modern study of the aging process— and geriatrics—the treatment of age-related diseases— also bore a burden. It was the burden of quackery. Early twentieth-century medicine, like late twentieth-century medicine, was filled with endless snake oil nostrums for aging and longevity. When McCay did his first CR experiments on rats, for example, entrepreneurs were still hawking rejuvenation compounds based on ground-up goat testicles and pituitaries from human cadavers. That legacy, rightly, birthed a culture of caution in the world of gerontology. You could see it whenever a professor or a scientist began talking "longevity" or "life span extension." Discussions of tenure suddenly grew terse. The invitations to fancy conferences dried up. The department chair—he was no longer introducing you to his lovely wife, who was, come to think of it, far too interested in all that rejuvenation talk.

Instead, the next thirty years saw gerontology focus almost completely on theories of aging. Specifically, why do we age? Among them were elaborations of the "rate of living" theory, itself proposed as early as 1928; it originally held that humans were born with a fixed amount of vital energy that was not replaceable and was burned up in the normal metabolic processes of living. (The theory's real value lay not in its restatement of Cornaro's radical moisture idea but in the connection between life span and metabolism.) There was the "mutation accumulation" theory (1952), which proposed an evolutionary basis for aging; here the British scholar Peter Medawar asserted that aging occurs because natural selection exerts its greatest force on the young organism—its only goal being reproduction—and, consequently, lets slide many unhealthy genetic mutations that do not show up until later in life. When organisms, from fruit flies to man, began living in safer environments (homes, glass tubes, cages) allowing them to avoid predation and to grow older chronologically, these deferred or accumulated mutations kicked in and cause what we call aging. A theory that pushed that observation further was known as "antagonistic pleiotropy," which made you sound smart just by pronouncing it; it asserted that the same evolu-

tionary adaptations that made a young organism fit to grow
and reproduce could become unhealthy later in life. Think, for
example, of a gene that allows you to store fat, thus surviv-
ing a famine but, in an environment of plenty, gives you type
2 diabetes. The "disposable soma" theory, another powerful
evolutionary model, held that, because organisms evolved to
favor early reproduction, they tended to apportion energy to
fertility and growth and shorted the maintenance and repair
needed later in life, hence, aging. One more theory is the
"glycation-glyco-oxidation thesis." An important permutation
of the old "rate of living," or "wear and tear" model, the "glycation thesis" held that the very process of glucose and protein
metabolism that takes place in normal cell tissue caused aging by creating stiff molecular "cross-links" in everything from
eye lenses to arteries. Many call it the "caramelization" theory,
referring to the observable manifestation of the theory one can
see when braising, say, a pot roast, although one might think
twice of telling dining companions that you are serving glyco-
lated bovine tissue tonight.

Two theories, however, served another unintended effect:
They pushed CR off the research agenda by the sheer force of
their proponents.

Perhaps the most *practically* influential theory arrived from
a man inspired not by rats or diet or aging or even biology,
but by basic chemical processes. His name was Denham Harman, the founder of the free radical theory of aging. Today, as
antioxidants are touted by everyone from Colgate Palmolive to
Coca-Cola, it is difficult to believe that there was ever a time
when science didn't "know" that free radicals—unpaired
electrons that cause damaging chain reactions with other

atoms and molecules—were the prime source of stress on the human cell and hence the engine of aging and disease. But the free radical theory of aging is a relatively new arrival. Harman came to it because, as he told it, he'd always been fascinated with the chemical side of medical science, and by chemistry in general. As a UC Berkeley undergraduate he spent several summers working for Shell Oil during World War II. There he landed in a fairly small department of the petroleum behemoth, the Reaction Kinetics Department, which studied reactive agents in organic chemicals. Reaction Kinetics wanted to put new compounds together and see if the reaction was commercially usable. One of Harman's discoveries would become the Shell No-Pest Strip. "I became interested in biology in part because I did not understand why some compounds with similar structures were effective, and others were not."

At about the same time, aging struck him as a possible specialty. An older student, thirty-one, applying to medical school, he was startled at being told, by admissions people, that he was already "too old—we want people who can practice for a long time." He recalls coming home one day, to his apartment in Berkeley, and seeing his vivacious eighty-year-old neighbor carry on with her business as if she were thirty years younger. He wondered: what really constitutes oldness? Gerontology, the medical subject, was in the air, as was life extension. Once, he recalled, his wife gave him an article on the subject in the *Ladies Home Journal,* titled "Tomorrow You May Be Younger." It was written by the science editor of the *New York Times.* "I thought it was extremely interesting and that chemists could be useful in medicine." Stanford Medical School, perhaps impressed with his pluck and drive, admitted him.

In 1954, Harman was working at the Donner Laboratory at UC Berkeley. He had just completed his MD, and, for a short period as a postdoc, he had a relatively undemanding schedule. He could, if he wanted, just sit at his desk and read journal articles all day. What he now knew about the human body began to come up against what he knew about chemistry. He contemplated aging as one primary flashpoint between the two. How did chemistry and aging interact? He began to make notes. As he did, his mind, as he recalled it, spun in two directions. Moving one way, he contemplated what he knew about the evolution of life, how it arose spontaneously 4 billion years ago when ionizing radiation from the primordial sun turned water and methane and hydrogen and ammonia into amino acids and nucleotides—the basic building blocks of life—through molecular and atomic reactions. Then he spun in the other direction: Everything eventually dies. Could there be a common mechanism? "I felt there had to be some common, some basic cause which is killing everything. We all go through this cycle of birth, aging and death. Everything. Not just you and I and other human beings. Bacteria, everything— nothing lasts forever. I figured there had to be some basic cause that was subject to genetic changes, because we all know that both genetics and environment have some influence."

If Harman had been at Berkeley ten years later, of course, he would have simply fired up another doobie and waited for all the heavy thinking to go away, but this was 1954, and a young postdoc with a young family had no such diverting options. There was just the older neighbor lady, who somehow was *aging really well,* and the experience from Shell Oil, where he'd had his first exposure to reactive molecules. Then, around

November 9, 1954, after four months of rumination and reading, Harman had an insight: "All of the sudden, the phrase
'free radicals' crossed my mind. You know, just out of the
blue. . . . And what I had was a profound sense of relief. I saw
something finally." He jotted it all down in his journal.

What Harman saw that day, and what he propounded
and reiterated over a fifty-year career (at this writing, he is
ninety-four and still publishing), was this: "Free radicals cause
random damage, and, depending on the type of radical, they
can cause all kinds of damage *from day one.*" Free radicals were
simply the unpaired, rogue electrons that formed when cells
burned fuel. They could cause the physical deterioration that
accompanies passing chronological time. Of course, it would
all be much more complex than that—there were all kinds of
agents the human cell produced on its own to counter such
damage (it makes its own antioxidants, for one)—but the basic
idea was there: When cells burn energy, their "exhaust" harms
human tissue, and that is aging. Through the 1960s Harman
documented the basic molecular reactions—the long, dreary
bench science that goes with all good theorizing—but there
it was, free radicals. In the early 1970s, Harman refined the
theory to take into account the free radicals produced by every
cell's powerhouse, the mitochondria, when they convert nutrients into ATP, the basic energy currency of cellular activity;
such free radicals could cause damage to DNA inside the mitochondria, and that could cause aging too—not to mention
cancer and all manner of mutation disorders. Harman showed
how consumption of antioxidants by lab animals increased life
span. An industry was born.

But Harman was not a great believer in caloric restriction.

Something about it did not feel right to him. He had nothing against expansion of maximum human life span. It was just that humans could not do so via CR, he reasoned, because, while lowered caloric intake likely led to reduced free radicals, it also led to less ATP, and that left humans far too hungry to voluntarily stick to such a diet. Instead, Harman focused almost exclusively on dietary antioxidant supplements. A number appeared promising and seemed to work in rodents. Humans have been taking them for decades now, with little evidence of their effect, either positive or negative. Harman still does not subscribe to CR.

Leonard Hayflick, a driven young microbiologist, conjured the other great aging theory of the postwar period. Hayflick was—and is—a blustery fellow, full of a sense of mission and proud of his penchant for "taking on orthodoxy," as he put it later. He was also talented, especially with the petri dish. His passion was bacteria in cell cultures. In the late 1950s, it was a red-hot discipline. Cell culture science—the ability to sample and propagate living cells in a medium of glucose and nutrients—had been around since the late nineteenth century, but as Hayflick rose through the ranks of young scientists at Pennsylvania's famed Wistar Institute, cell culture techniques grew ever more refined. The search for vaccines, which used those techniques extensively, flourished.

In that pursuit, Hayflick tripped on a central scientific dogma of the period. This was the belief that all in vitro cells— cells grown under glass—were, technically, immortal. If they were kept in the correct nutritional and environmental conditions, they would continue to divide and repopulate ad infinitum. Hayflick, playing around with various microbacteria, detected a flaw in that theory when he began actually counting the number of times a population of cells doubled. There seemed to be a limit—fifty doublings—no matter how good the conditions were. He tried to publish the results, but no one believed him. Then, in an act of laboratory chutzpah that would become legend, Hayflick played a kind of scientific trick: to all of the doubters of his findings—some of the biggest players in cell biology—he sent copies of the same line of cells and simply asked the recipient to telephone him when and *if* the line began to die after fifty divisions. Every single one grudgingly did, and so was born what is now known as the Hayflick limit. "I knew I had something then."

But why did the cells stop dividing? Fortunately for Hayflick, by now the normal human chromosome number had been determined, and cytogenetics—the ability to type a cell's gene patterns—was rising. Hayflick had his research partner perform the analysis, and the finding was remarkable. As Hayflick recalled, "The cells had stopped dividing not because of any accident or ignorance about the [petri dish] culture media, but because of some internal clock." That internal clock, he would eventually deduce, came from the slow but steady wearing away, cell division after cell division, of the ends of chromosomes, dubbed telomeres. Telomeres are a kind of protective cap to chromosomes, and, as Hayflick's theory evolved, came to be viewed as a kind of timekeeper to healthy aging. If you could somehow prevent their erosion, you would stanch the inevitable molecular disharmony that flowed from an uncapped chromosome—aka, disease, impairment, death. Telomere aging came to be viewed as a cell's internal meter, a signpost on its way to death.

For cell biologists studying aging, the Hayflick limit and telomere shortening were powerful theoretical tools. They helped shift much of the search for causes of aging from extra-cellular events—things that happened outside the body—to intracellular events, things that happened in the cell itself. It was a profound change, but it also led to some serious dead ends. Hayflick's observations were, after all, in vitro—in glass—not in vivo—in a living organism. Also: not all tissue cells in the human body divide—the heart and the brain, for example—but the heart and the brain did age. How did you explain that? There were huge holes in the theory; although the limit was upheld in many in vitro tests, there were also lots of cells that did not stop dividing at fifty; some stopped at two hundred; others showed little sign of ever slowing down.

There was also the philosophical impact of Hayflick*ism*. If, as Hayflick told it, all cells were fated to age and die, how could anyone in their right mind propose any way to extend maximum life span or slow aging? "We wrongly assume that aging is a disease," he liked to say. "Aging is just a series of changes that make the body more susceptible to disease." In this, Hayflick became the scientific ideologue for much of postwar gerontology. If saying that aging was a disease repre-sented ageism to the social scientists and "gray power" activ-ists, it was a scientific heresy to Hayflick, a throwback to "the dark arts," as he put it. If he closed his eyes and imagined what the world would look like if a pill existed to end aging, or even to slow it down? Why, he would explain, *it was not pretty.* The rich would not only dominate economically, but demographi-cally. Tyrants, he wrote, would live forever. And, perhaps most symbolic of his generation, Hayflick believed that a longevity pill would deprive people of the "joys of aging," of following the sun in one's RV and not having to worry about the kids. And the kids! He wondered: What if children decided to age normally while parents did not? "Humans' ability to tamper with the aging process will produce societal dislocations and effects on human institutions that will be monumental."

Of course, almost every one of Len Hayflick's critiques rang with some truth, but to some scientists interested in longevity, Hayflick's limit began to feel, well, limiting. "I don't know why Len and that gang are so intent on stopping people from talking about things like life span extension and other strategies," said Steve Spindler, a cell biologist working on CR at UC Riverside. "It's like he had one good idea and spent the rest of his life defending it. What good does it do?" It all began to grate. "I found it to be an unworthy hypothesis. Studying cells in a culture cannot give you an organismal [or whole organism] understanding," said Edward Masoro, a professor of physiology at the University of Texas. "Hayflick was going around implying that his theory was, essentially, aging under glass. I rejected it as a hypothesis."

In science, of course, it is one thing to reject someone else's hypothesis, and quite another to propose and test one of your own.

Of mice and men and rats

Edward Masoro evoked the quintessence of modern laboratory science. Or at least the popular image of it. A short, compact man with a bristly crew cut and horn-rimmed glasses that dominated his scrunched-up visage, Masoro was a master of mammalian physiology—specifically of the rat and rat metabolism. He knew everything about *rattus,* and with such specificity that he once wrote an entire book about how the little rodent metabolized fats. He had done his graduate work in physiology at UC Berkeley in the 1950s, and by the late 1960s, when he first began encountering "gerontology types" at various scientific conferences, he had finally found a home at the University of Texas in San Antonio. There, he became known for his ever-evolving courses on mammalian physiology, and for his excitable teaching style; Masoro once got so caught up in a lecture about rat metabolism that he fell off the stage while talking. At the end of a lecture, people were always unwinding him from his microphone cord. In other forums, he had no qualms about openly challenging colleagues on their theories. His pet peeve was often framed by some form of the exclamation: "But that has almost nothing to do with the *actual animal's physiology!*"

The gerontology bug bit when Masoro heard a man named Morris Ross talk about caloric restriction and the retardation of aging. He was impressed with the careful quality of Ross's work, and he grew intrigued with the subject—or, more precisely, the general ignorance of the subject. "The more I heard, the more it became apparent to me that neither I nor anyone else knew anything about aging." He also liked the fact that he might bring a fresh perspective—that of metabolism—to what seemed like a settled, or at least ignored, subject. "If you went to the Gerontology Society of America conventions and talked caloric restriction, everyone put it down because 'everybody knew' that it was from delayed development, and of course you couldn't do that to people," Masoro recalled. "And all the nutritionists 'knew' that the CR effect was from the loss of fat."

To deduce exactly how CR worked to extend life span, Masoro devised a meticulous series of experiments to test a variety of proposed CR mechanisms. They were not the kind of sexy, big-picture experimental pieces that lay readers of twenty-first-century science have come to expect, but, rather, the tedious, step-by-step work of a master laboratory scientist whose research only gets looked at when everyone else is trying to figure out how to design their own experiments. First Masoro set out to define how a normal rat ages: How did skeletal muscle change with age? What happened to fat deposits with aging? How did aging affect the rat's digestive process and breathing? Having established normal aging parameters for the animal, he then began a series of classic caloric restriction experiments. One group of rats in any given experiment would get to eat as much as they wanted—usually referred to as *ad lib*—and one group would be restricted to 60 percent of their usual intake. A key component of the regimen was to ensure that the restricted mice received all essential nutrients and vitamins, the goal always being "undernutrition without malnutrition." They were, in a sense, Cornaro rats.

Right away, the tests began revealing all kinds of health and longevity benefits. The rats on CR not only displayed a longer average life span—by up to 40 percent—but they also had a longer maximum life span compared with controls. This was a fundamental point. As McCay had noted forty-five years before, CR was not altering life expectancy, it was altering something fundamental—maximum life span, thought to be fixed. CR in Masoro's rats also consistently delayed or prevented the onset of almost all of the classic diseases of aging, from kidney and liver disease to heart disease and sarcopenia, or loss of muscle mass. CR prevented the age-related decline in insulin sensitivity and glucose metabolism, the cause of adult-onset diabetes.

In rapid succession, Masoro's tests destroyed the old assumptions about how and why CR extended life span. Was it because CR reduced an animal's overall metabolic rate—the number of calories burned per unit of body mass—thereby reducing wear and tear on the body? No. Instead, Masoro showed that "the food-restricted rats consumed a greater number of calories per gram of body weight during their lifetimes than did the rats fed *ad lib*, yet they lived longer." (It was hard here not to think of one of Cornaro's more pithy epithets: "*To eat for long, eat little.*") Was it because CR, as McCay had held, interrupted their growth and somehow slowed down the aging process? No. Masoro showed this by initiating CR in adult rats, then compared them with controls and with rats restricted right after weaning. The results were unequivocal: The adult restricted rats benefited almost as much as the rats restricted from youth. Was it because the CR rats lost weight? No, in fact, the CR rats that lost the least weight lived the longest. Was it because of some dietary change—in the percentage of fat or protein or minerals in their feed? No. Again, the key finding was that CR's life-span-extending effects—and one couldn't help but recall Cornaro's most basic injunction—came purely from reduced calories, or energy, consumed during the lifetime of the animal. It wasn't what they ate, it was how much.

So, if it wasn't any of that, what was it? Between the lines, Masoro detected two consistent trends: One was CR's impact on sugar metabolism and insulin signaling. Animals on CR showed little of the traditional decline in insulin sensitivity associated with aging. They maintained much lower blood sugar counts. Perhaps as a result of that, they also showed far less glyco-oxidation in vital tissues—the "carmelization" that led to eye disease and arterial stiffening. But what was the mechanism behind that? A growing suspect was stress. Somehow, Masoro and his colleague Steven Austad theorized, the mild daily stress of CR triggered what biologists have dubbed "hormesis"—a beneficial action from something usually considered detrimental. In this case, the mild stress from a voluntary lack of food shifts an organism's energy use to repair and maintenance of tissue and away from growth and reproduction—a kind of famine response—thereby slowing down the onset of aging. *It was all about how the body used fuel.* Some called it "energetics." As if to further underscore the stress thesis, Masoro showed that CR was consistently associated with moderate daily *increases* in stress hormones called glucocorticoids.

As Masoro digested his data, he came to some interesting metaconclusions, something he generally didn't do. The more he looked at aging, disease, and CR, the more he came to disdain the old ideological dividing line over "aging as a disease versus aging as a distinct and discrete process." At bottom, it was all totally arbitrary, he said. After all, a neoplasia, or tumor growth, can also be seen as a loss of homeostasis, or physiological balance. Bone loss is a universal process in all mammals, but when it becomes intense, it is called osteoporosis. "It is totally arbitrary," Masoro said. But is *calling* aging a disease a form of ageism? He had no truck with that line, either. "Aging is something that is bad. There is no such thing as successful aging. It is all aging and it is all bad. You cannot mix up science and sociology."

But science and sociology *are* fundamentally linked, especially in a society where medicine and consumerism have fused. The question for Masoro soon became: What do you *do* with all that CR data? On that, Masoro remained an intransigent rodent physiologist. You could not extrapolate it to humans. "I always thought CR was a bad idea for humans," he said. "Tell them to do that, and they might do anything."

Roy Walford, the era's other great figure in CR research, had few qualms about caloric restriction for humans. Like Masoro, he too had parsed CR's various effects and mechanisms. But there any resemblance ended.

Walford was an imposing figure, with a Fu Manchu mustache and a perennially shaved head long before it was fashionable. He was constantly traveling to faraway places. Once, he traversed some of the remotest regions of India, carrying with him a bag full of rectal thermometers and temperature gauges to measure the subnormal temperatures of long-living mountain yogis. He was outwardly political, covering the 1968 Paris student riots for the *Los Angeles Free Press.* He had a penchant for theatre—he was a member of a mime troupe—and for art—he painted and sketched and even did video art. He rode a motorcycle and wore a leather jacket and could be seen, any given Sunday, roller-skating along the

Venice Beach boardwalk, not far from where he lived, his bald pate shining in the bright California sun.

Walford was obsessed with aging and death. From an early age, he took to quoting from the story of Faust; his daughter, Lisa, recalls that "anyone who knew my father knew what the central theme was: Faust, and the unfair bargain between life and death." As early as 1941, at age seventeen, Walford was protesting it all. In an article in his high school's literary magazine he titled "Conquest of the Future," he complained that "elders have received positively no gain from science concerning expectant life span . . . but death is not a necessary adjunct of living matter." It was hardly surprising that he became a physician and a research pathologist. He wanted to know why people got old and died—and how it could be stopped, or substantially delayed.

His research palate mirrored his personality. Walford's core interest lay in the role of the immune system in aging, but he seemed to know few bounds for exploring the topic. If Masoro was married to the rat, Walford was always dating new and exotic research species. There was his work on "life span, chronologic disease patterns, and age-related changes in relative spleen weights for the Mongolian gerbil"; his articles about the "effect of temperature-transfer on growth of laboratory populations of a South American annual fish *Cynolebias bellottii*"; and, of course, the family-fun favorite, "life span and lymphoma-incidence of mice injected at birth with spleen cells across a weak histocompatibility locus." As a longtime colleague put it, "Roy was intensely creative in his explorations—he didn't know limits—or at least he was not afraid of them."

Beyond the imposing outward demeanor—perhaps made more imposing by his subdued manner—one other thing was clear: Roy Walford was an outstanding scientist. In a brilliant series of experiments using genetically uniform mice, Walford, usually teamed with fellow UCLA professor Richard Weindruch, fleshed out a broad range of CR effects: how, even when begun in mouse adulthood, CR slowed down or prevented age-related cancers; how it prevented the loss of immune function; how it preserved liver cells and even increased the ability of such cells to function; how it slowed down the loss of gamma crystallens in the mouse eye lens, a key cause of age-related vision problems in humans as well; how CR improved learning and motor skills in aged mice; and, strangely, how it failed to prevent age-related neurochemical buildup. The more Walford looked, the more he saw the connection to human aging. It was exciting. In 1982, writing in the journal *Clinical Geriatric Medicine* about life extension in mice, he took the proverbial leap out of his old Faustian universe: "With a fairly high order of probability, the same might be obtained in humans. There is no reason to insist that maximum life span in humans is irretrievably fixed."

He took up CR himself, and began experimenting—cautiously—with vitamin supplements. He assembled an "optimal nutrient" diet and menu for humans who practiced CR, including, among other things, "the perfect CR muffin." He wrote a book titled *Maximum Life Span*. He wrote another called *The 120-Year Diet*. There were experiments on himself and friends; he practiced a one-day fast, one-day moderate eating regimen on CR. (One of his best friends, the prominent USC gerontologist Caleb Finch, once joked that "whenever

I saw Roy, it just so happened that *that* day was his eating day.") His consumption of marijuana was legendary, which might be justified because of his interest in its hypothermic, antiaging effects, but one hopes that it was mainly because he had a lot of fun with it. He immersed himself in yoga. All of this Walford did while maintaining a vigorous, competitive laboratory known worldwide for creativity and scientific rigor. Increasingly he cleaved to "systems biology," the study of large-scale interrelationships and ecological dynamics. And so, in 1992, it was hardly surprising that, when a rich Texas oilman proposed to build a self-sustaining sealed environment called Biosphere 2 in the Arizona desert, Roy Walford would be one of the "terranauts." To Walford, Biosphere 2 was akin to going on Darwin's *Beagle.* You did not know what you were going to find out, he later said, "because nature was going to ask the questions."

Although huge sums of money were spent on the three-acre complex to make it self-sustaining, Biosphere 2 never really worked. Life inside was often nasty, brutish, and bitchy. Coaxing the environment to produce enough food for the crew of eight required seventy-hour workweeks, and even then it was not enough. The crew was forced to practice CR. Worse, Biosphere's artificial lung, down in its vast steel-and-concrete basement, did not produce enough air; the crew was chronically airsick, or hypoxic. They lost weight fast, something Walford knew was bad—it released a flood of toxins from quickly shrinking fat cells. No one was having a particularly good time, despite the media's portrayal of it as a kind of Shangri-la in the desert.

When the crew emerged in 1994, it soon became clear that

something had changed in Roy Walford. He seemed . . . wound up. "They started calling him 'the barking Pekinese of med sciences,' " a former UCLA colleague recalls. At science conferences, he was antagonistic to those who disagreed with him. "It was very unpleasant when we would meet, because I inevitably was the guy the media called to question whether CR should be done by humans, and of course I didn't agree with him. He was always very angry and confrontational with me," recalls Ed Masoro. "We ended up hating each other." A few years later, Walford himself began noticing other changes. His way of walking seemed stilted, labored. He grew disoriented. Eventually he got himself checked out. The diagnosis was devastating. He had all the symptoms of ALS, a form of Parkinsonism known as Lou Gehrig's disease.

In the four years preceding his death, Walford vigorously milked the Biosphere data for everything he could. It had been, after all, a completely unplanned CR experiment on humans. Studying the crew's blood tests, he found that the low-fat, nutrient-dense diet in Biosphere 2 "significantly lowered blood glucose, total leukocyte count, cholesterol, and blood pressure." Just as in his mice. He also detected strange blood reactions to the chronic hypoxia, or low oxygen levels, that the crew experienced; their blood was akin to that displayed by animals that go into hibernation. This, he theorized, may have been because of the crew's low-calorie diet. Walford was so absorbed by the Biosphere data, and produced so much from it, that, even after he died in 2004 of respiratory failure, publications continued to print new work by him. One of his last articles was in the journal *Movement Disorders.* Its concern was reflected in its title, "Atypical Parkinsonism and Motor Neuron Syndrome in a Biosphere 2 Participant: A Possible Complication of Chronic Hypoxia and Carbon Monoxide Toxicity?"

The next year, at the CR Society conference in Tucson, Arizona, a number of members drove out to see Biosphere 2, the experiment that may have killed the man who paved their way to a longer, healthier life.

They found that it had been transformed into a tourist attraction.

Of love and sex and the CR longevity phenotype

The next time I saw Michael Rae, he was not spouting "rubbish" to the speaker at the podium, but, rather, embracing his girlfriend, April Smith, also a CR practitioner. We were at CR Three, held in the small banquet room of a Mexican restaurant in downtown San Antonio. Smith had just given a presentation about how the media had come to portray the CR Society—negatively—and she was intent on spinning it all the other way. She took to talking about one of her favorite subjects—Michael Rae. "I mean, I was attracted to Michael for a very basic, superficial reason," she told the audience. "I think skinny guys are just so hot. But the more I got into CR, and Michael's experience of it, the more I saw it had a lot of unexpected benefits. Like the end of a monthly menstrual period—what is wrong with that? And also something that Michael talks to me about all the time, which is that, before CR, when he met a girl, the sum total of his thoughts were, you know, 'I wonder if she'll sleep with me? I want to have sex with her. Will she have sex with me? Sex with her would be great I bet.' And ad infinitum. But that, after CR, his thinking is totally different. More along the lines of, you know, 'she might be an interesting person.' There's no comparison, you know." Had Rae shifted himself into a state of hormesis, when the low but constant voluntary stress of CR makes you less likely to copulate and more likely to age slowly?

The audience in the banquet room roused. These are the kinds of things CR people love to hear about, because they are bits of a puzzle about whether the science of CR in mice matches the experience of CR in people. It is a kind of mouse-o-centrism, or, for lack of a better term, "mousomorphism." Mice on CR have greatly reduced fertility, which seems to track with most CR practitioners' experience of reduced libido; it is a basic evolutionary, or life history, trade-off. ("It is just not that important anymore," one longtime restrictor told me.) Mice on CR have slightly impaired wound healing, which seems to hold true for many CR people, although usually only in the beginning of their restriction. The mice have a lowered body temperature; ditto CR people. CR people often complain about a sensitive bum because of the lack of padding down there; mice do not complain. One of the things that does *not* seem to track, as far as I have seen, is outward display of energy. A calorie-restricted mouse at two years of age, compared with a non-CR mouse of the same age, displays tons of movement and energy. CR people, who claim huge boosts in energy reserves, do not show it. If anything, they are, as Lisa Walford told me, dampened down.

David Fisher, a British member of CR, presents with a classic case. I had met him in Arizona and had found him, as most

Americans find the British, charming and engaged, even a lit-tle wry. But that was only after hanging around him for some time. To the outsider he would appear just as Lisa Walford said—dampened down. And, like so many members of CR, he evoked a slightly naive quality when it came to discussing his early awareness of mortality and aging. "I became aware of aging and mortality as a child and began to dread them. I was always fascinated by science and assumed that one day aging would be cured, but that this may be in hundreds of years, per-haps when the body could be rebuilt molecule by molecule." He went on: Sometime during the late 1980s, he read about McCay's experiments and, seeing that the aging process was more immediately malleable, took up CR. "People overem-phasize how hard it is," he told me when asked about it. "It's only the first five years that are uncomfortable." He did not smile when he said that. He currently practices a kind of cave-man CR, he said. Lots of nuts and berries and some fish and meat. He is about fifty-three—and looks, perhaps, five years the younger, albeit a somewhat strained five years younger.

There is also the sense among some that CR is a refuge from a world spinning out of control—a place where one can feel some sense of certainty. One man in his thirties, who I will call Kevin because he does not like to appear in the media, ex-plained his conversion to CR this way: "It eventually dawned on me that if I could live long enough for the technology to be developed, I could escape this life of quiet desperation in a way that didn't involve dying. I could have my new brain pro-grammed to avoid, for example, libido—a source of much suf-fering; and I could have time to gather enough resources to not be dependent on, say, having a job. Thus I decided to try a little bit harder to eat less. I still never said, 'OK, now I'm on CR.' I just tried to prepare the most nutrient dense foods I could, and then eat less of them." So the religious mind comes into play as well. Of overt religiosity there is little among CR people, although Michael Rae practices a form of ancient, bare-bones Christianity that requires dedication to a larger mission in life.

Mainly the cult of science prevails. Stick with any CR conference long enough and you will likely hear about every objection to CR that you can imagine—this because the CR Society seems to invite debate and challenge. It seemed to me a healthy inclination, one that, say, the AMA might try occasionally. On just one afternoon the Society members were told, among other things, that "the CR peace dividend," or benefit in humans, will be small—six to seven years at most—*if* you extrapolated from the mouse data to two known human caloric extremes, Okinawan war survivors and Sumo wrestlers, *then* plotted that line on a graphic of caloric intake per unit of body weight (to which the main murmured response was, "I'll *take* it!"); that CR's much-vaunted effect on free radical damage might actually not matter very much when it comes to actual aging rates, which is sort of like being told you really didn't have to eat all that spinach in the first place; that studies of CR members' carotid arteries showed a much slower progression of arterial stiffening than in non-CR people, which had everybody vaguely fingering their throats and smiling; that the growing knowledge of how mice and men differed on CR— CR people, for example, do not show reduced IGF-1 signaling, while CR mice did—was "a troubling finding," as John Holloszy, perhaps the dean of human CR research, said; and that gene studies of livers from mice that started CR very late in life showed activation of several known extended life span genes—which made me look around for Michael Rae and April Smith, to see if they had heard that and run out to gorge on chips and boink like rabbits. To all of that, the audience listened intently, quietly, and then followed with detailed, nuanced, and utterly bloodless questions. They wanted to know the truth. It was impressive.

But once in a while, a strange thing happens among some CR folk. They begin to think that, besides standing in for lab animals, they can also stand in for scientists.

I caught a glimpse of this during an endless weekend spent at the Tarrytown Sheraton in the summer of 2007. I was there at the invitation of Paul McGlothin, the chief scientific officer of the CR Society, and his wife, Meredith Averill. The meeting, McGlothin told me, would be about "the future of CR." I had a hint of the agenda. McGlothin had been sending out flyers on the Internet, advertising a "Glucose Control Workshop." He sent me some suggested reading references as well. The list was heavy on two basic ideas: hormesis, and the growing body of knowledge linking it with insulin signaling. The topic was doubly hot because two scientists, Leonard Guarente and David Sinclair, had shown that CR triggered the gene product sirtuin, which led to enhanced tissue-maintenance, and, consequently, life span extension. In yeast. The pair had also made a compound from red wine, called resveratrol, that seemed to do what CR does. It seemed to work for obese mice on a high-fat diet, a model—let's face it—for a third of the U.S. population. Maybe one could get the life extension action without giving up food and sex and a round ass. So the talk in the CR community was all about "CR mimetics"—of the

long hoped-for CR pill. All of this had led McGlothin to pro-
claim to me, in one triumphal phone call, that "the future of
CR is all about cell signaling!" That the science is far from the
try-this-at-home conclusiveness that the FDA, let alone your
own doctor, might accept? McGlothin had an answer: "You
just have to find the right doctor!"

A former professional clarinet player turned ad man,
McGlothin, whose voice and elastic features conjure a kind
of low-key Don Knotts, had been practicing CR for fourteen
years when, he said, he'd grown weary with "the old image,
you know, the guy in the dark corner office with his wheat
germ who'll live a hundred years and hate every minute of it."
As he spoke, he began to physically "perk himself up" in front
of me, as though to prove what he was saying. "That's not us!
That's not *me*! I mean, I run a superedgy ad agency, superedgy,
and a lot of my employees practice CR too. I am positive that
a fifty-nine-year-old guy like me could never be performing so
well without it. And that's the way with most people in CR.
The image is out of whack, big-time!" (As David Harrison, one
of the leading mouse longevity researchers, later noted to me
on my blog, "At 59, he *should* have energy, whether he is on CR
or not! He is only 59!") There was something stilted and practiced, poorly, about McGlothin's little diatribe, but I let it go.

As if to further prove his point, McGlothin had thrown
the workshop net wide to recruit new members. It seemed to
have worked. Although about half of the attendees were classic, pencil-thin restrictors, including charming David Fisher
and dark libido "Kevin," the rest were Joe and Jane Averages,
several with the requisite doughy American physiognomy.
Two had outright paunches. They were newcomers. There

was a tall fellow named Eddy from Brooklyn, who was there because he had "just finished chemo, and I'm gonna rebuild myself from the ground up, and this seems like a way to do it." There was a gauzily clad woman named Julia from Seattle who had suffered a "total digestive collapse" after picking up a parasite in India. And there was a tiny expatriate Sony executive named Dave from Tokyo who explained, as he ate his dinner, gram by gram from a portable scale, that "I just had a close relative die of cancer, and if there's anything I can do to prevent it happening to me, then so be it. I've been doing this for a year now and thought I'd come to get a full indoctrination of state-of-the-art CR."

On Saturday morning, the indoctrination, McGlothin style, commenced. He focused on two elements. The first was what he called tight glycemic control. "You've got to control your blood sugar and insulin," he said, encouraging everyone at the conference table to pick up their new (drug-company-provided) glucose meters, which came in their conference goody bags. "You've got to keep your blood sugar from slamming *up*"—he tapped the desk and pointed to a chart projected on a screen— "and *down*! You've got to keep it in a narrow range, because if you don't, you start sending a message to your body to make too much of a lot of bad actors, things like IGF-1 and TNF, things that are markers of aging and chronic disease." And as if to leave the Atkins option slammed closed, he advocated limiting the amount of meat we eat. It was vegans, after all, who had the lowest IGF-1 counts.

It was morning, and breakfast (boiled sweet potatoes and green bean mash and fruit) awaited, but before "feasting," McGlothin insisted that everyone first take a blood sugar reading to get a "baseline." This resulted in lots of fumbling with the finger-pricking equipment ("Oww! Three hits and still none. I must be anemic!" "Fuck! Baseline *this*!") After having everyone record the number that appeared on their meters, McGlothin instructed the group to eat a "tease meal"—a few chunks of sweet potato only—to provoke an insulin response; the idea was to have your insulin up and ready to "smooth out" your blood sugar when you finally eat your "real" breakfast a while later.

Then, just as several attendees looked as if they were about to sprinkle salt and pepper on their chair cushions and dig in, McGlothin and Averill popped up out of their seats. "Time for a walk," he chirped, doffing a floppy canvas hat. "It's a way to make sure you're setting yourself up for good cell signaling. Sometimes Merrill just jumps rope with weights on her back for a few minutes, but we can just walk." Everyone walked, then ate, kind of.

Lifestyle makes up the second element of McGlothin's "New CR Way." We were shown the correct way to prepare for sleep—at least three hours in dim light before sleeping—meditation techniques, supplements that raise cognitive performance. There was a long discussion of resveratrol, and, totally straight, "how do you do your own home blood testing?" But none of us was quite prepared for what McGlothin proclaimed as "the next phase" of CR's evolution. After the Sunday morning breakfast "tease," we found out. "I call it the CR daily fast," McGlothin proudly told a few slightly puzzled newbies, one of whom muttered, "I thought we were *already* fasting!" The essence of it, McGlothin went on, was simple: People should consider going for a one-hour walk *instead* of dinner every day.

That, he said as he displayed another round of slides, would re-sult in all kinds of positive cell signaling benefits, ranging from memory improvements to, of course, better glucose control. "Merrill and I got the idea a few years ago, when we went for a long walk after our main meal at lunchtime and then just didn't eat the last meal of the day. It felt great. And we sort of looked at each other and said, gee, you mean for forty years we ate dinner instead of doing this? Why did we do that?"

The room was silent.

McGlothin and Averill beamed idiotically.

Suddenly, I couldn't take it anymore. Because you are an effing human being! I wanted to yell. Sitting there, I won-dered if McGlothin had gone completely bonkers, if someone had stolen a couple of cards from his deck, if his elevator didn't go to the top floor anymore. Perhaps it was because I knew that a number of the things he was saying were just wrong, or so equivocal as to be so. He was recommending a lowered-*protein*, as well as low-cal and low-carb, diet. This he justified on the basis that protein restriction was found to be better than caloric restriction in reducing insulin-like growth factor 1 levels in Chinese famine survivors, which *everybody* just *knew* was the way to go because transgenic mice with low IGF-1 live the longest. Besides the fact that such a regimen might make you wish you *were* in a Chinese famine, the advice runs counter to just about every single CR experi-ment to date. People, like regular mice, need adequate and regular protein; it is better to eat a little more than too little of it. Similarly, humans, *unlike* mice, need IGF-1 for a broad range of things, maintaining heart and brain cells, for one, and fighting infection and injury, for another. In fact, in the limited human data that exists on CR in humans, hadn't we learned that IGF-1 is *not* lowered? What did he have to say about that? Why, McGlothin didn't even drink wine, the divine milk of the aged, for God's sake. I caught him by the arm on the way back from another tease-walk, or whatever, and asked him about that. "It's not worth it," he said, his rubbery countenance a little tighter. "I had a neighbor who practiced CR. His only vice was booze. He died at age seventy-nine from pancreatic cancer."

Oh.

Now McGlothin brought the room to a hush and asked everyone to take off their shoes to "feel rooted in now." He explained how meditation was the fourth new component of his CR Way, and then led everyone, eyes shut, in a guided visualization of their internal body. "And now, let's just thank our pancreas," he intoned solemnly, "because with the tease, the walk, and the meditation, it has been producing insulin and getting it to the right level." He went on to the other organs. Each . . . and . . . every . . . organ.

Sitting there, thanking my colon, I couldn't help but wonder if this is what Roy Walford, a cosmopolitan Renaissance man with a great sense of conviviality, had in mind when he started CR. It couldn't be. Had Ed Masoro been right when he declared that CR was not for people, who "if you tell them to do that . . . they'll do anything"? I thanked my left ventricle and rubbed my eyes, praying for an escape. There was silence.

Then, from outside the conference room, came sounds of life. The non-CR world had roused itself. Food platters banged and juice glasses tinkled. The smell of bacon wafted about. At the Sheraton Tarrytown, Sunday brunch had commenced.

"Morty?" drifted in one Brooklyn voice. "Do you want the French toast or what?"

And that, really, is the question, isn't it? Do you want the French toast, or not? Do you want your extended life to be a life, or not? There had to be a better way than the cold way. The hungry way. The flat-ass no-sex way. And sure enough, American medicine and American business were cooking up something really hot.

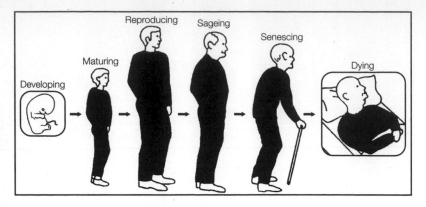

The Six Stages of the Life Course

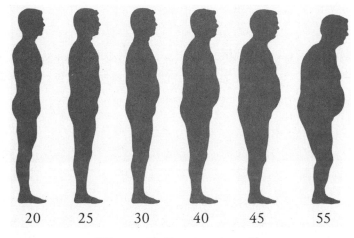

| 20 | 25 | 30 | 40 | 45 | 55 |

Silhouette de l'homme régressif

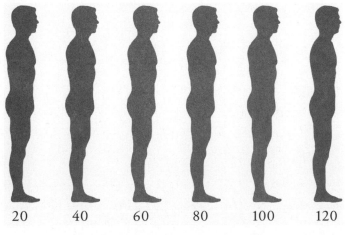

| 20 | 40 | 60 | 80 | 100 | 120 |

Silhouette de l'homme progressif

BOOK TWO

*C*ash

*Well, now, Doctor, just in confidence I'm going to
tell you something that may strike you as funny, but
I believe that foxes' lungs are fine for asthma, and
T.B. too. I told that to a Sioux City pulmonary specialist one time and he laughed at me — said it wasn't
scientific — and I said to him, "Hell! Scientific!" I
said, "I don't know if it's the latest fad or wrinkle
in science or not," I said, "but I get results and that's
what I'm looking fir's results!" I said.*

— SINCLAIR LEWIS, *ARROWSMITH* (1925)

One way to wrap your head around the gulf between today's
brand of commercial, antiaging medicine and the world of
academic longevity science is to look at the graphics to the left.
The first, entitled "The Six Stages of the Life Course," comes
from a university textbook, *The Molecular Biology of Human
Aging*; it represents the classic, humanistic impulse of modern
gerontology to define various stages in the aging process and

vest them with some meaning. The second—a pair of original illustrations—was inspired by drawings found in a book by a man named Georges Debled, almost unknown to American doctors but who, for some time, dominated the popular European discussion of male aging and the use of testosterone. The two are not even in the same universe.

It was easy to come by the first graphic, reflecting as it does our modern neoclassical ideas about aging—that it is "natural" and "inevitable." It was difficult to come by the second. It was only after traveling for some time in the world of hormones and antiaging medicine that it came to my attention. As a male, it held special appeal. Not that I am *that* fat, but that, at the age of fifty-three, I had a testosterone problem; I didn't have any. It had taken a while to discover that, and to find out why. I had suffered a moderate concussion in a horseback riding accident a few years ago, and, ever since, I had displayed all the signs of an aging man: extreme grumpiness, becoming overly startled at sudden noises, becoming easily overwhelmed, a decline in libido, and an overall loss of energy. Not for nothing has hypogonadism been dubbed a form of accelerated aging; and not surprising is the fact that, if you search the catalog of the National Library of Medicine with the words "testosterone" and "aging," you get some of the few journal articles in existence with the word "grumpiness" as a medical term. Mainstream endocrinology recognizes the syndrome and cites it as one reason all concussion patients should get tested for hormone levels. Yet in all the top-flight, big university med school consultations that followed my case, no one—no one—ever suggested, let alone prescribed, such a test. I was left wondering if I would be a proto-senior forever. Hormone replacement, the single most powerful element in the modern antiaging apothecary, loomed. What was it all about?

"Rich guys playing with their hormones," said Dr. Fran Kaufman, one of the foremost endocrinologists in the United States. "You know, basically, that's what you're talking about." Kaufman and her husband, Dr. Neal Kaufman, and I were sitting in a West Los Angeles Starbucks on a rainy Sunday afternoon, talking hormones, aging, and health. Fran Kaufman had agreed to look over some PowerPoint presentations that I had been given by Dr. Ron Rothenberg, a prominent antiaging physician; I had interviewed Kaufman before on a number of articles and had come to appreciate her depth of knowledge, openness to new ideas, and worldly sense of humor. And so we had flicked through the PowerPoints, one on testosterone, one on human growth hormone, and one on estrogen. As I tapped the computer pad, I noted, out of the corner of my eye, something odd. Fran Kaufman, usually hyperattentive, seemed, frankly . . . a little bored.

I pointed out the data supporting hormone supplementation. She didn't rise to it. I asked her if I should try it. Ho-hum. What about growth hormone?

Neal, a respected clinician and one of L.A.'s more agile thinkers about health care, jumped in. He had his own ideas about hormone supplementation. "No one talks about the fact that perhaps nature—or evolution, whatever—intended for humans to have low hormones as they age. Think about it: low estrogen and low sex drive after giving birth makes sense for women; their priorities and role in life have changed. The same with growth hormone; once we are past the time of life when we grow, physiologically . . . well, how much do you need? And testosterone—you know, *that* has a huge cancer risk, and— ”

"But Neal, that's not true," Fran broke in.

"It's not?"

"No, not testosterone. That's not the big issue with that. You can give it to guys who want it, as long as you have the right monitoring and lab tests. It can be a huge help, too. The issue there is almost like giving thyroid—the issue is, can they *stand* it? A lot of guys get it and don't like the way it feels, at least until a week or so after the injection. And frankly, if it helps them, how much should we care about what evolution intended?"

"Well," Neal said, taking a sip of his coffee. "I'm still not sure. *I* wouldn't prescribe."

We laughed. Neal's penchant to consider the vast complexity of health care is the subject of endless discussion, and Fran often gibes him about it. "Yes, I know, but *when* would you ever prescribe *any*thing!" We all laughed again.

Driving home, back up through the rain-soaked Sepulveda Pass, I began to wonder: Was hormone supplementation for me?

The Great Grasela

Once upon a time, hormones—from the Greek *hormao*, meaning "impetus"—held a place of pride in American medicine, both for specific disease treatment and for longevity enhancement. Testosterone provides the perfect example. The principal male sex hormone, T, as it is abbreviated, is made in the testes after it is stimulated by follicle stimulating hormone, or FSH, itself made in the pituitary. Its origin in the pituitary, a small, saddle-shaped gland that sits behind the base of the brain, explains its connection to concussion; if you're out long enough, or get hit hard enough, you get impaired secretion of FSH, and, consequently, low testosterone. (This is why football players and boxers always seem so grumpy.) That's what happened to me, but to a less dramatic degree it is what happens to all men after the age of about forty.

Since the late nineteenth century, medicine has been trying to find a way to "top off" lowered T in fellows like me. In 1889, the Harvard professor Charles-Édouard Brown-Séquard reported his "rejuvenating" injections of dog and guinea pig testicle extract in the *Lancet*. His experiments were not easily replicable, and when they were, the treatment was found to be weak and transient at best. There followed two decades of what might be called circus medicine—testicles from corpses sewn into the thighs of prisoners, which produced very crabby prisoners but little else; extracts of racehorse balls and tinctures of monkey pee, which lightened the pockets of the nation, but did little else; and any number of strange and bizarre testes transplants in chickens, roosters, pigs, goats, horses, and, of course, mice and rats. The consequence of this oft-misguided research was predictable, especially in America, where commercial medical science seems only to have an "on" and an "off" button. Brown-Séquard died in ridicule, as did any reputable research on male sex hormones; that is, until the 1930s, when two things happened.

One was that the basic physiology of hormones got worked

out by the young field of endocrinology. It goes like this: Specific chemical messengers secreted by specific glands are carried by the blood to "targets" or receptors on cell surfaces. The messages then get transcribed by the inner cell, which in turn sends a reverse signal to the originating gland, telling it to stop secretion. Such is the essence of the now well-established endocrine concept of a negative feedback loop. The other breakthrough came in manufacturing, as large pharmaceutical companies developed methods for cheaply converting cow cholesterol into chemical testosterone. (A parallel process of the era succeeded in deriving estrogen and progesterone from wild yams and soybeans.) The first wave of therapeutic use, in the 1940s and 1950s, was followed by scandal, as the chemical methyl-testosterone was found to be responsible for a number of cancers. The consequence—played out in the realms of other hormones as well—was reluctance by clinicians to prescribe it, even long after safe versions of T came on market. Robert Butler, the dean of modern gerontology and no friend of today's antiaging medical crowd, recalls trying, with no success, to get the NIH to sponsor clinical trials of testosterone therapy on older men. "They were just too frightened of the cancer risk."

Bubbling underneath the medical establishment, however, two powerful trends were unfolding that were less driven by fear and more driven by opportunity. The first was the rebirth, phoenixlike, of the traditional compounding pharmacy—the kind once depicted in Norman Rockwell paintings. In an era of Walmart and mail-order prescriptions, it is hard to believe that pharmacy compounding, the on-site mixing of raw chemicals in accord with a written prescription from a physician, had been the norm, worldwide, until the 1940s, when large drug companies began to standardize doses, usually in tablet or capsule form. By the 1980s, 99 percent of retail pharmacies simply counted out prepackaged doses into little containers. The art of compounding—and, frankly, some of its inexactitudes—was lost, but so was the ability of modern medicine to respond, simply, to the vast range of dosage variances inherent in as diverse a population as that of the United States.

Then something happened. That something was about "hard-ons," says John Grasela, now one of the leading compounders in the United States. Or, rather, "the inability of older men to get sex-enhancing drugs out of the traditional drug system." I had met Grasela at the annual convention of the American Academy of Anti-Aging Medicine (A4M), where his pharmacy, University Compounding Pharmacy (UCP), was sponsoring a session entitled "How to Turn Your Practice into a Cash-Based Anti-aging Business Now!" His booth was mobbed, and when I asked for an interview, he suggested I drive down to San Diego to talk to him the next week. "I can't get loose now!" I took him up on it and met up with him at UCP headquarters, a low-slung commercial building just on the outskirts of downtown San Diego.

Grasela is a tall, somewhat taciturn-looking man in his early fifties, with the demeanor of a small businessman—all purpose—and a tan. Over a cup of hazelnut coffee and his "daily dose" of some twenty vitamins and supplements, he gave me a look at "the future" of the pharmacy in America. "This is for when your dick gets crooked," he explained, gesturing to a tube of ointment for a compound that had been discontinued by one of the large drug companies some time ago. "And this is where we mix the testosterone cream," he said, tapping on a big metal bowl and mixer. "We can do five pounds at a time!" Trained as a pharmacist at Wayne State, Grasela and his partner sold off their first business venture, a chain of drugstores in Detroit in the late 1980s and went west. "We basically took the money and ran," he says of the midlife career move. "We had a license in California and we worked our way down the coast, looking for an opportunity. We bought our first store in San Diego and we realized what we'd walked into was an aging population. These were people who came here during World War II and worked in the naval yards and retired." These were also people who wanted to keep screwing. Grasela started getting calls for "stuff" the drug companies had dropped because of low sales, like paparavine, a vasodilator that could be used "for hard-ons," as Grasela puts it. Then there was pentolamine, another hard-o-nergic blood pressure med also dropped for low sales. "It was a great opportunity to get started in compounding, and it got me interested in aging." He then went to one of a small number of antiaging physicians in the area for a personal consultation. There, he recalls, he got the hormone gospel: "It was a revelation. I'd never thought about aging, and about replacing only what your body needs. That made sense to me." Soon, he was on a full regimen of hormones and vitamins. He felt fantastic—"leaner, with more energy, and that's not to even mention the creative juices!"

But why, he kept thinking, weren't more physicians practicing antiaging medicine? The short answer was that hormones are complicated, they didn't always work as intended, and to prescribe them responsibly required a physician to order a lot of detailed blood tests, for which insurance companies and managed care did not want to pay. It was a stalemate. Then Grasela got a brainstorm. Fuck the insurance companies! (It was a sentiment, shall we say, that was shared by many.) The real issue was education. If you could educate average clinicians about how to diagnose, treat, and monitor hormonal deficiencies in older people, and find a way to standardize the lab tests to ensure people did not get sick and sue you, you could transform your practice into a cash practice. Patients—the right ones—would pay cash for antiaging medicine. Patients. Cash. Antiaging.

So Grasela did what many pharmaceutical companies do when they want to prepare the way for an expensive new drug. He recruited "thought leaders," or experts, in various fields of hormone replacement to give paid workshops on the subject. He advertised in trade publications and underwrote health shows on cable TV. The workshops were an instant hit. "The physicians are basically being driven into it by their patients, who are very motivated patients, cash patients," he says. "And that allows the physicians to reinvent their practice and gets them out of the pharmaceutical corner. They like it for another reason too. It is fun medicine." Fun? "Fun in that it is not about trying to get a guy to diet and exercise." He cited a recent Las Vegas workshop as an example of "how mainstream" hormones have become. "Of the 270 doctors there, 80 percent were rookies to a hormone workshop." His own business has been transformed as well. At his state-of-the-art compounding operation alone, Grasela now has twenty-eight full-time pharmacists; he fills more than five hundred prescriptions a day. "But I have not lost sight that this is a fun business," he says. "This is not about keeping people from dying. We want to leave at five thirty every day. It's quality of life."

The offices of Dr. Ron Rothenberg, one of Grasela's thought leaders, sits off the San Diego freeway, not far from the famed Scripps Medical Center. Rothenberg had agreed to speak to me about antiaging medicine, and to treat my hypogonadism. I had heard him talk at an antiaging medical convention on the topic and had read his book *Forever Ageless*. He seemed sane, if, to my mind, a bit overoptimistic. His offices, which are modern, spacious but not particularly lavish, conjured a slightly oceanic feeling, with pastel colors and paintings of fish all around. Despite this, I did not hate Rothenberg. I found him to be a pleasantly engaged fellow with a refreshingly *nebbische surfische* sensibility—one underscored by the red, white, and blue surfboard hanging on his office wall, and the Hawaiian shirt under his clinician whites.

Coming out of a traditional medical education, Rothenberg, like a number of antiaging physicians, originally specialized in emergency medicine. "It made sense for a guy like me, who has a kind of hands-on approach to life in general," he said. When he got a little older—and started seeing the signs of aging in himself—he started asking questions. Mainly about hormones. "Why hadn't we been educated about them more?" he says. He began reading the various theories of aging and, in the not-unreasonable amalgam of understanding that he calls "the ongoing search for the truth as a moving target," came to believe that the underlying cause of aging is hormonal. "We age because our hormones decline, not vice versa," he likes to say. Although there are deep, deep problems with this statement—what, for example, causes the thymus to wither, the pituitary to stop making growth hormone, or the adrenals to stop making cortisol?—it is not off the grid either;

the Stanford evolutionary biologist Robert Sapolsky has advocated something similar in many of the most respected journals in the world. Testosterone decline—apart from regular age-related drops in testosterone—is also hot among epidemiologists, with one large study showing that a sixty-five-year-old in 2002 had lower testosterone levels than a sixty-five-year-old in 1987. No one knows why, but the highly flaccid chart illustrating that drop . . .

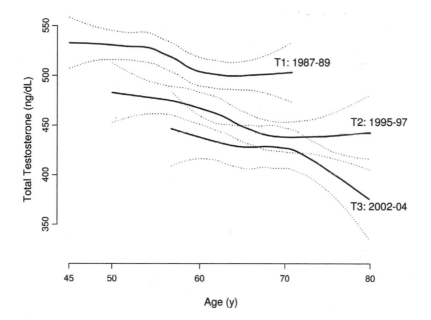

. . . is one of Rothenberg's favorite talking points.

What separates Rothenberg from the academics is his willingness to prescribe hormonal replacement for levels that the establishment deems "within the reference range," or normal, or even "low normal." "What is low normal—what is that?" he says with a Shecky Greene shrug. "Using that philosophy, do you only give a person who is nearsighted glasses that make them see as they might have at forty, rather than at twenty? What is that?" Applied to hormones, in my case testosterone, that means bringing my low-normal score of 350 (the range is 300–1000) up to youthful levels. He looks at my lab tests and then looks over at me, eyes sparkling, and chuckles. "Oh, you are going to feel so much better!"

And, truth be told, poorer. For if anything, the cost of the lab tests ($500, cash, to be repeated at least twice a year) presages the other most notable difference between Rothenberg and my usual family physician: the utterance of the word "money." When Rothenberg's office first received my lab tests, his assistant was on the phone, asking me if I wanted the basic lifestyle plan ($3,000) or the advanced antiaging plan ($12,000). A lot of this depended upon which hormones I decided to take, but there's an interesting social contract at work as well, one usually not articulated in traditional clinical medicine. It is this: You pay me x per month, or I will not approve the refill of your prescription. Not surprisingly, I took the minimum—Rothenberg insisted that I take a thyroid extract as well as T because I was "low normal" on that—and I aggressively refused the lifestyle consultations on food and exercise.

Nevertheless, the "sell" of supplements—Rothenberg says, and I believe, he does not make money off the items but rather offers them at his wholesale price—can be off-putting. As I was waiting for him to perform a physical, two female assistants came into the examination room to give me samples of some kind of "acacia blueberry antioxidant" drink. I asked them to leave. They looked . . . hurt. Later, the same pair loaded me up with free samples of various blender drink supplements. They gave me a long lecture about the right time of the morning to drink this concoction. One of them then gave me a cursory lecture about warning signs from the testosterone cream I would be getting—from, of course, University Compounding. "Basically," she said, "you can't use too much of this stuff. But if your nipples start to tingle, dis . . . con . . . *tin* . . . ue! And, oh yeah, don't smear it on your testicles." Put it on your arms, thighs, or chest. Oh . . . yeah.

The money, combined with the nontraditional yet scientifically fortified presentation, would be off-putting to anyone, but I wondered how off-putting much of traditional medical treatment for chronic disease would sound if the terms were stated just as baldly, instead of hidden behind the baloney boilerplate of HMO-speak. (Prozac as "first line treatment for clinical depression" might be translated as "we will pay for Prozac because Lilly will give us a good deal for buying it in bulk, but not for therapy, because your therapist will not work for $5 an hour.") I decided to cut Rothenberg a break on that matter—there are plenty of ethics police out there already, *there you go boys; go get 'em!*—and simply adopt a pragmatic attitude: Would it work? Clearly there were some measurable ways to assess this antiaging regimen. My cholesterol was high—and had remained so despite an energetic lifestyle and a reasonable, regular Mediterranean diet, albeit one supplemented with cookies and milk. We could check that in six months and $500 or so. My blood sugar and insulin were OK by conventional measures, but Rothenberg thought they could be a lot better. "You're flirting with diabetes." We could measure that, too. And, of course, there were the persistent unmeasureables that had pushed me toward treatment in the first place: the jumpiness, the lack of energy, and what I called "cognitive swamping." Would I get better? We'd see.

Before leaving, Rothenberg asked me in for one more mini-consult. "You know," he said, "your growth hormone levels are low, too—you qualify for supplementation of HGH, you know." But isn't that dangerous? "Are you kidding?" he said. "That's just the official line. If you go over there to Scripps and ask the big shots about it, they'll all tell you, 'don't do it, it might cause cancer,' and all that. But let me tell you—half of them are on it themselves, just like me. And they love it."

Growth does not equal death—or does it?

HGH—short for human growth hormone—is the second force that freed vast swaths of clinical medicine from its anti-hormone bias. In short: it was just too good not to try it.

Growth hormone, made by the pituitary, has often been called the master hormone. Its effects on the human body run broad and deep. A protein comprised of 191 amino acids, its main job is to signal the liver and other organs to produce insulin-like growth factor 1, or IGF-1, the bogeyman molecule that Paul McGlothlin at the CR Society hates so much. As its name suggests, IGF-1 is responsible for growth and maturation of bones, muscles, and cartilage. IGF-1 is also a potent anti-infection hormone; it repairs and replaces destroyed cell tissues, and it protects the heart. Just how HGH makes IGF-I, and how IGF-1 functions both on the whole body and on its individual parts, may be one of the most complex single investigations in all of medical science today. No one knows exactly how the pathway works, though there are a number of good beginnings.

This relative dearth of science, of course, never stopped anyone from, as you might guess, *grinding up the pituitaries of dead people* and injecting said pituitaries into children with growth hormone deficiencies. Perhaps the only thing that prevented mass grave robberies was the singular invention, by Genentech in 1981, of a way to synthesize vast quantities of HGH using modern recombinant DNA technologies. By 1985, somatropin was on the market, approved by the FDA for use only on children with documented deficits in growth hormone. Lilly, Pfizer, and the rest followed with their own versions. Of course, everyone over forty or so runs low on growth hormone; as Neal Kaufman alluded, evolutionary biologists long believed this was an adaption to long life spans—lack of growth hormone might be an evolutionary check on cancer, a form of unhealthy growth that happens more and more frequently as people live longer. Although that theory has been broadly challenged in recent years, it was one reason that many clinicians were wary of using the hormone, even when clearly warranted in borderline dwarf children.

Then, in 1990, a physician and medical researcher at the University of Minnesota, Daniel Rudman, had a breakthrough. Rudman was interested in how elderly men, displaying all the signs of normal aging, might respond to HGH therapy. It was not a theoretical pursuit. Rudman had a long and abiding interest in how frailty in the elderly could be treated, or, better, prevented. For six months, three times a week, he had twelve "overly healthy" (not fat and not too skinny) men aged sixty to eighty years inject themselves with the hormone; these men, like an estimated one-third of all men in that age group, had less than 350 U/liter of blood of IGF-1, indicating a substantial lack of growth hormone. (Healthy men between twenty and forty average somewhere between 500 and 1500 U.) Nine other subjects in the control group received nothing. All followed the same diet.

After six months, the results were striking. Although there was no substantial weight gain in either group, the men in Group One experienced two profound changes: an 8.8 percent increase in lean body mass, and a 14.4 percent loss of adipose tissue—exactly the opposite trend displayed in almost all aging human bodies. Shrinking livers and kidneys got bigger. The men also registered substantial thickening of their aged skin, and small but important gains in bone mass. Although Rudman had many reservations about the intervention, he could not help but phrase the findings in a way that would be ... noted. "The effects of six months of human growth hormone on lean body mass and adipose-tissue mass," he wrote, "were equivalent in magnitude to the changes incurred during 10 to 20 years of aging."

Here we pause for a test on American medical culture. Finish the following sentence:

"Quickly the findings were seized upon by (choose one) ..."

(a) the gerontology community, which tried to convince the government to conduct wider trials in hopes of adding something to the thin pharmacopeia for dealing with frailty, a growing cause of death among the aged.

(b) the AMA, which was concerned, as the population aged, that there existed no good way to deal with sarcopenia, the muscle wasting that accompanies old age and which is responsible for many deaths that result from falls by the elderly.

(c) pharmaceutical companies, which, seeing that there were not enough dwarves around anymore to buy their HGH, decided to hype the findings.

(d) two "bone doctors" from Chicago, one of whom held the *Guinness Book* world record for one-armed push-ups and handstands, who wrote a best-selling book promising that HGH could "stop the clock" on aging.

Bone doctors, or osteopaths, have been discriminated against for decades in the United States, although, like compounding pharmacists, they once held considerable sway. The problem is that osteopaths are largely practitioners of holistic medicine, and in a society that demands scientific evidence to justify a treatment, theirs has been a vocation of faith—faith with a few footnotes. Nevertheless, their numbers have begun to climb back up in recent years, fueled in no small part by older Americans who've had a gut full of modern medicine and pharmaceuticals. DOs now occupy a fast-evolving niche of medicine, fast because they tend to be more experimental than their brethren MDs.

In the 1980s, few were more enterprising than Robert Goldman and Ronald Klatz, two Chicago bone doctors with an interest in sports medicine. Goldman, a weight lifter with broad shoulders, a trim waist, and curly hair, was fascinated with steroids; Klatz, the Poindexter of the pair, was an inventor, constantly coming up with new patents for brain-cooling

devices and bladder catheters. They both eventually specialized in sports medicine. After Klatz suffered a bad car accident in 1991, the pair gravitated toward the subject of aging,
and in meetings, first informal, later more organized, they
began to talk about aging as a disease. "It was very exciting,
because, here we all were, in the middle of careers, and we
began seeing things differently," recalls Vincent Giampapa,
a plastic surgeon who hung out with the pair. "The idea that
aging was a disease, and not something that was the natural
course of things, was radical, but we saw the connections."
Convening a like-minded group in the global medical science
capital of Cancun, Mexico, in 1992, they formed the American Academy of Anti-Aging Medicine, often shortened as
A4M. They bought a large, neo-Gothic mansion in Chicago
for headquarters.

Right from the beginning, growth hormone promotion
loomed large in A4M's evolution. As Klatz tells the story,
"Also present at the [Cancun] conference was another group
of people, who were not part of the scientific crowd. They
were a striking bunch, trim, muscular, sexy, energetic, upbeat,
with the vibrant look of good health. Yet these were people in
their fifties, sixties, even seventies. They talked about changes
they had undergone—melting away of fat, increased muscle
tone even in the absence of exercise, a vastly increased sense
of well-being." They were in Cancun because it was the only
place they could get treatment with growth hormone. "Then
a strange thing happened," Klatz recalled. "The doctors and
researchers who had assembled to form this organization devoted to aging were suddenly on the defensive. Could these
claims be true? Had these people truly stepped back in time,

or were they suffering a mass delusion? 'My God,' said one of the scientists, 'we're all talking about the possibility of reversing the aging process and here these people are actually doing it.' " Within a few years, Klatz and Goldman were HGH converts. The FDA had by then been persuaded to broaden the definition of who could be prescribed HGH. In 1996, the pair published *Stopping the Clock: Dramatic Breakthroughs in Anti-Aging and Age Reversal Techniques;* the next year, Klatz published *Grow Young with HGH.*

Should you take it? The claims in both books were largely based on anecdotal case reports and small studies of people with severely abnormal growth hormone deficiencies, not the moderate deficits experienced by most aging people. But long-term studies would be several years in coming, Klatz argued. And even then, who knew how long it would take the medical establishment to sign on to HGH replacement? It had taken forever for them to adopt estrogen and progesterone replacement for women. "We believe the consequences of not acting are far worse than the consequences of acting," he wrote. Every day that goes by, the aging process is eroding vital capacities "in every cell in our bodies." Deciding whether or not to go on HGH could have life or death consequences for all. Waiting for definitive studies was not worth it. It was time to act.

There was one other extremely controversial claim. It was this: that growth hormone and other replacement therapies would add thirty years to maximum life span.

Now everyone was interested in HGH. Sales soared. Celebrities endorsed it. The visage of Dr. Jeffry Life, the sixty-eight-year-old proponent of hormone replacement for antiaging, graced the pages of the nation's toniest magazines. "Only replace what you lose naturally"—the great Grasela's maxim—rang though the land, especially in retirement communities. It "just made sense."

The illustrations on the office walls of Andrzej Bartke—one of a mouse, one of a sunflower that Bartke drew himself, and one showing mortality curves of mice put on caloric restriction— suggest the kind of grand, expansive, and eccentric world that all great naturalists seem to crave. One thinks of Darwin or Huxley, perhaps even Linnaeus, the great classifier of species, or Gesner, the Renaissance master of zoology. It is a world of complexity, nuance, doubt, wonder—and more doubt. It is not, to put it mildly, the kind of world that embraces ideas like "only replace what you lose naturally" because it "just makes sense." This is exactly why Andrzej Bartke, a courtly, Krakow-born endocrinologist with a fondness for zoology, has become public enemy number one to many in today's commercial antiaging medicine community. "Has the guy *ever* treated a patient?"—the ultimate put-down in that go-ahead world—is how Ron Rothenberg described him and other critics. Bartke's sin is that he has dared to demonstrate two deeply troubling physiological facts in mice: one, that continuous, high levels of human growth hormone *accelerate* aging, and two, that low levels of it may *extend* maximum life span. In describing the results of his work, he is fond of one qualifying caveat: "in mice." Or, as he uncharacteristically emphasized in his normally emotion-free e-mails to me: "*IN MICE!*" Yet despite his persistent caveats, Bartke's work nevertheless has made it into all kinds of public forums about human aging. In 2007, Bartke's research was even entered into congressional testimony about HGH use by professional baseball players, who are not mice.

As Bartke, a professor at Southern Illinois University School of Medicine, tells it, he got into the subject almost accidentally. Until the early 1990s, his work centered on a completely different hormone system, that of prolactin, a stress hormone that Bartke discovered played an important role in fertility. He studied it in two strains of mutant, dwarf mice—the Snell and the Ames—that carry a mutated pituitary gland, and, hence, little or none of the hormones signaled by that gland. "It was kind of an endocrinologist's dream," he said in his mild-mannered, almost monkish way of discoursing, head bowed, holding his brow with one hand. "I mean, you had one condition, and you had one hormone that could treat the condition." It was a dream because things endocrinological—the endless systems of signals, negative feedback loops, and cell surface receptors that go into hormonal science—just did not unwind that way very often. So Bartke burrowed in further. He was on to something.

But while Bartke, in test after test, cautiously characterized the role of prolactin in mice, his fellows at other institutions were onto a much sexier pursuit—genetically engineering human growth hormone into cows, pigs, mice, and other animals. Much of the work was funded by the government, with an eye, ultimately, on increasing meat production in cows and pigs. One of Bartke's friends and mentors, Tom Wagner, at Ohio State, succeeded in developing a strain of transgenic, or TG, mice that overexpressed growth hormone. The result was remarkable, if, in retrospect, predictable: animals that were twice to three times as large as normal mice, promptly dubbed "giant mice." Wagner and others were just as interested in the scientific technique—at the time, introducing a foreign gene to an organism by using an activating third gene—but an unforeseen threat to the giants' ability to reproduce brought them Bartke's way. The giants—they were infertile.

Fairly quickly, Bartke figured out what was in play. The overexpression of growth hormone had tweaked other hormonal systems, in particular that of prolactin. "We knew we could fix that by giving the mice prolactin," he says. "But the thing that got all of our attention"—meaning his colleagues Holly and Kurt Brown-Borg and himself—"was the condition of the animals. They looked awful, like they were falling apart. You could see it when you compared them to normal mice— the TG mice had the humped-up back, the fur loss, all of it— and they tended to die off sooner. That got our attention. When we dissected, they had aged kidneys and other glands. What could it be?" Eventually, Holly Brown-Borg put up a counterquestion. "Well," she said to Bartke, "*you* worked with the dwarves"—the Snells and the Ames mutants that had the opposite problem, little or no growth hormone—"how long do *they* live?" As was the case with McCay and his rats sixty years before, the answer was "I don't know," Bartke says. "We had always euthanized when the experiment was done." For the next three years, the trio ran experiments comparing normal and Ames dwarf mice, which lack the pituitary gene known as Pit1. It was difficult getting funding, "because everyone was on the growth hormone track and really did not want to hear about it, everybody was talking about Rudman." Eventually a small grant came, tellingly, from the USDA.

The results of the Bartke trial, published in *Nature* in 1996, were clear and unequivocating: The dwarf mice lived 50 percent longer than normal mice. Not only that, they, much like CR mice, had far lower rates of almost all diseases, including cancer. They also showed delayed aging of the brain and arteries—fewer inflammogens, plaques, and glycation products. The theoretical implications were far-reaching; for the first time, a single gene mutation in a mouse could be shown to extend life span, thus placing the mouse and mammals in the same arena of life-extending genes as the fruit fly, the worm, and yeast. All showed major life span extensions when IGF-1, the hormone generated by growth hormone, was down-regulated, or reduced. As Richard Miller, of the University of Michigan Geriatrics Center and one of the world's foremost scholars on mice and aging, noted, "this was a remarkable piece of work." It held the promise of fundamentally changing our understanding of how mammalian aging—and by some extension, human life span—were controlled by specific genes. Predictably, leaders in the antiaging community were incensed. "They wrote this incredibly detailed, passionate letter to the editors of *Nature*," Bartke recalled. "It was . . . disturbing in that it basically said how dare someone imply that low growth hormone and reduced IGF-1 is good." He gently spread his arms, smiled, and shrugged. What can you do?

But the accusation, and the barrage of data that the anti-aging proponents of HGH submitted to support their case, got Bartke's attention. He noted that almost everything Goldman, Klatz, and others had written referred to "more than 2,000" studies that proved their case, that HGH was a bona fide anti-aging agent. Bartke began checking their data, and what he found disturbed him. "Almost every one of these supposedly solid studies was based on case studies of patients who were middle aged, and who had vastly abnormal growth hormone levels, not elderly people with normal levels, which you might have expected since they were saying that HGH was an anti-aging drug for the elderly."

Bartke and his colleagues went on, as did Masoro and Walford, to parse and refine their own results. What was it about reduced—*but not completely blocked*—growth hormone and IGF-1 that extended life span? He put Snell and Ames dwarves, which make little or no growth hormone—or thyroid, for that matter, because of their pituitary mutation—on caloric restriction. The result was further increased life span. He then tried CR on a mouse with a different mutation, one that generated its own growth hormone, but which lacks a complete receptor for it on the cell surface, leading to completely blocked IGF-1 action. That mouse, named GHRKO for growth hormone receptor knock out, showed no further life span extension from CR. To Bartke, that just showed how complex the hormonal connection to life span extension could be. There were other hormones involved, he saw, and growth hormone itself was probably doing other things in the body that we do not understand. And IGF-I, undoubtedly important to heart and bone and metabolic health, was likely being autonomously generated by other organs, like the heart and brain. "And remember," he likes to say, "this is in mice that we are showing this, not people, although it would be very unusual for a trait that seems preserved in everything from yeast to mouse not to have some bearing on humans and human aging."

But there was an important corollary to CR humans: the dwarf mice were infertile and cold, the GHRKO mice were

not. All of this implied some complicated connection to metabolism, reproduction, and multiple hormones—not just
ones that you could easily "top off." As Masoro had noted
years before, it all seemed to connect to the way fuel was used
in the body, the way it was partitioned. The dwarves and the
GHRKO seemed to partition fuel to be used for maintenance
and repair of the body, the giants seemed to partition it to favor
rapid and excessive growth, and the normal mice somewhere
in between. The implications for aging grew clearer, especially
if you believed, as do many gerontologists, that a huge part of
the underlying aging process is the loss of the ability to repair
damage to cells.

There was something else. In almost all of Bartke's scenarios
of extended life span, on the one hand, and shortened life span,
on the other, the great constant was insulin sensitivity. The
dwarves maintain it to the end, as do mice on CR; the giants,
while they have it early on, rapidly begin to lose insulin sensitivity, leading to diabetes and accelerated aging. Why the dispar-
ity? One emerging thesis is fat cell patterning. Dwarf mice are
obese, but almost all of their fat is the more beneficial subcutaneous fat. Subcutaneous fat cells are smaller, and secrete tons
of a hormone called adiponectin, which raises insulin sensitivity as well as anti-inflammatory and anti-atherogenic factors.
Bartke wonders at the consistency: "What strikes me about all
this is that what we are seeing in these long-lived mice is the
opposite of metabolic syndrome," the obesity-related conflux
of high blood pressure, insulin resistance, and high cholesterol
seen in a growing percentage of the population. "Having an
opposite syndrome is a huge advantage."

I asked Bartke if this was his way of saying that enhanced

insulin sensitivity, rather than stress resistance, ought to be the Holy Grail of life extension study. He smiled beneficently. We had been talking about the exciting work emerging from the laboratories of Leonard Guarente and David Sinclair, the resveratrol proponents whose work on sirtuins has tended to emphasize stress resistance to oxidative damage as its main life-span-extending effect. We had also been talking about how the oxidative damage theory of aging had not exactly been faring well in recent years, with a number of studies showing that, in terms of aging, oxidative damage may be a relatively minor player. Sinclair, after all, had sold his company to Glaxo for $750 million last year, mainly because early clinical tests demonstrated that his compound had potential as a diabetes drug; a later study of resveratrol in normal, as opposed to high-fat-eating mice, showed it had no life-span-extending properties. Bartke seemed to have anticipated that himself a few years ago when, upon concluding one article, he wrote, "We suspect that research efforts to develop 'CR mimetics' for pharmaceutical intervention in the aging process may be more effective if they focus on targets in the GH/IGF-1/insulin signaling axis."

What are you trying to say? I asked again. He sheepishly handed me a piece of paper. On it was a flowchart that detailed various theories of aging and how they intertwined. At its core loomed the words "decreased growth hormone." From that box flowed all manner of boxes and arrows, from "decreased cancer incidence" to "increased stress resistance." The only point on this map that had *three* distinct intersecting lines was a little bubble marked with the words "increased insulin sensitivity."

"That's what I'm trying to say," Bartke said, nodding gently at the diagram. He smiled even more beneficently. "In *mice!*"

If Bartke has been slow to extrapolate his results to humans, a vocal core of scholars in the gerontology establishment—Big G—has not. The gerontologist Jay Olshansky, from the University of Chicago's School of Public Health, took a leading role in the anti-growth-hormone crusade in mid-2005, proffering Bartke's work as his standby. Olshansky, a big, bearish man, and his friend, the quieter Dr. Thomas Perls, who oversees the New England Centenarian Study, were so convinced that all antiaging medicine was quackery that they began saying so in a variety of scholarly forums, which led to their being sued for libel by Klatz and Goldman. The case was later settled, but it left both scholars even more vigorously opposed to growth hormone. Perls took the case to Congress, citing Bartke's work—something one of his peers told me was "intellectually dishonest," and another characterized as "unproductive and, let's face it, where does it end when you start censoring people for their views?" (I later had the chance to ask Perls if he regretted the campaign, to which he replied, "I only wish we could get a prosecutor who would be willing to pursue a RICO case against these guys!") Olshansky managed to marshal a broad group of scholars to issue a "consensus statement" on the "fallacy" of antiaging medicine. When I finally got a chance to ask Olshansky exactly what it was that he found so objectionable, he sat me down and lectured me. "That anyone can sit there and actually use the words 'antiaging' and 'science' is beyond me," he said. "Anyone who runs one of those clinics will be in jail in five years. They will all be shut down. You heard it here first!" I asked him what it was that got him so motivated. "I wouldn't have anything against it if it were not for the fact that the chief source of information about this for the public seems to be some TV celebrity," referring to Suzanne Somers, who wrote two lucrative best sellers on the subject.

Meanwhile, the actual scientific data about humans, longevity, health, and human growth hormone—be it pro or con—has remained puzzling, murky, tainted with agendaism and strange gaps of basic research. Perhaps the most negative and influential piece of scholarship about HGH and humans came out of Stanford University in 2007. It was not a clinical trial but rather a review of older works. The conclusion: "Claims that growth hormone enhances physical performance are not supported by the scientific literature. Although the limited available evidence suggests that growth hormone increases lean body mass, it may not improve strength; in addition, it may worsen exercise capacity and increase adverse events. More research is needed to conclusively determine the effects of growth hormone on athletic performance." But just as Bartke noted that Klatz and Goldman's pro-HGH studies were mainly based on treatment of middle-aged patients with very low levels of GH, much of Stanford's study looked at superhigh, or supraphysiological, doses in professional athletes, not older people taking low-dose HGH, as is the norm in antiaging practices. One widely circulated anti-HGH study showed an increase in type 2 diabetes in users of HGH, but that was in morbidly obese users who were administered large doses. Those given low, frequent dosing did not experience the complication. Similarly, the legitimate fear of cancer—cancer, after all, is unchecked growth, and growth hormone can fuel it—has yet to be sussed out with hard data. Standard protocol in antiaging practices—and one reason it costs so much—is for tests that monitor early cancer indications.

Epidemiologically, there are huge gaps. Acromegaly—which results in facial disfiguring and a range of physiological problems—is the best-established disease consequent of too-high levels of growth hormone, but its rate of occurrence has remained constant over the twenty years since HGH was introduced. Similarly, the few studies looking at growth hormone expression in centenarians have been troubled by methodological problems. A much-vaunted study of 384 Ashkenazy centenarians indicated that only nine carried a gene for disrupted growth hormone transmission—the human equivalent of the GHRKO mouse. That minuscule result has left even anti-HGH scholars scratching their heads. "How did that ever get published?" more than one told me.

One reason for the rush-to-negative judgment is that, in a nation where health care is increasingly a consumer product, driven by ads on TV, perhaps a default that errs on the side of caution is a good thing. (It would be even better if the same ethos applied to almost all chronic disease drugs.) It might be different if the antiaging profession stepped up to the plate and did its own oversight of adverse events, what the FDA calls pharmaco vigilance. But every time I mentioned this prospect—to everyone from Grasela to the head of the International Pharmacy Compounding Association—I was told that it would be impractical and expensive. That leaves the debate in the hands of static interests—vested economic interests in the case of the antiaging industry, and vested intellectual, cultural, and political interests in the case of the academics.

"There is a huge and legitimate scientific debate that everyone is just dancing around on this," says Steven Austad, an evolutionary biologist studying aging at the University of Texas Health Science Center, and one of the scholars who signed one of Olshansky's anti-antiaging salvos. "The fact is that growth hormone serves a much broader and deeper purpose in humans than it does in mice and flies and worms. It serves heart and brain health. And, whether anyone out there studying human dwarves and centenarians likes to admit it or not, humans with low growth hormone do not do well. They are prone to all kinds of medical problems. I wonder sometimes how miserable they are. On the other hand, there is no data that HGH taken from [a needle] is going to extend your maximum life span, and the cancer risk is still unclear."

I decided to goad Rothenberg on this point, and because I was due for a six-month checkup of my testosterone and thyroid replacement, drove down to his offices in Encinitas. I asked him: How long does an average client stay on growth hormone? This is an important, basic question, tracked relentlessly in other drugs by pharma because it is the best single initial indicator of patient satisfaction and the drug's effectiveness; if the patient is not experiencing benefits, the patient does not renew the prescription. With HGH running upward of $1,000 a month, that one piece of data might tell us a lot. Rothenberg hemmed and hawed. He didn't want to make generalizations. What about himself? "I'm back on it—like a lot of people, I go off and on as I feel the need," he said. So what is really the controlling factor? "Let's face it," he said. "For a lot of people, it is the money. It's expensive. For some the effects are modest. So a lot of it is about how much money you want to spend on modest improvements." But what about life span extension, the great claim by Klatz and Goldman? "We have to face it—antiaging has a name problem. That's why I have that name, and not antiaging, on my door."

Before I left, we reviewed my own hormone supplementation. I had discontinued the thyroid because it made me feel nauseous and agitated, but I had kept up with the testosterone. I felt a small elevation of my spirits; I could cut back on my dreaded antidepressants, and my waist had shrunk by at least an inch. But Rothenberg had a surprise. With a twinkle in his eye, he pulled out my most recent lab tests. For the first time in my adult life my cholesterol, which runs high despite a good diet and exercise, was almost acceptable—a total just above 200. Previously it had always run between 240 and 270—Lipitor country. Although my blood sugar has always been decent, my new tests showed that it, too, had improved. It was enough to make me suspend judgment just a little longer and wonder about what Bartke wondered: What about all those other hormones? How did they affect aging, and will supplementation slow it down?

The return of the early twentieth century

If you wanted to find out about hormones and their place in the contemporary antiaging firmament, you'd want to go to the annual conference of the American Association of Anti-Aging

Medicine, A4M, held each year at the opulently tacky Venetian Hotel in Las Vegas. It is, of course, a circus; one wishes for a Twain or a Mencken, or perhaps even a David Sedaris, to vet its quintessentially American nature. The A4M public relations people, sensing the barrel of fish they have as a client, severely restrict press access, but with a little patience, they can be convinced that you will give their fish an even break. Once in the barrel, it is difficult to do so. Here, laid out on a ballroom floor the size of the Gator Bowl, with a good size sprinkling of "demo girls," "pavilion hostesses," and, let's face it, outright working girls just to make it *really* resemble an AMA convention, clamor purveyors of every known antiaging product under the sun.

A sampler might include the Sunetics Corporation's "Laser Hair Brush," about which the company touts: "Fact: the FDA cleared the first Laser Hair Therapy device to grow hair in January 2007," although one wondered what happened in February. There was LivOn Laboratories' Lypo-Spheric AGE Blocker, "the most powerful oral nutrient delivery system in the world," competing, apparently, with the human mouth. There was the "Barefoot connections" antiaging device sold by Earthing Solutions, which helps the earth's electric field "transfer easily to the body," as if it had problems doing so in the first place. One physician panelist who had her own booth gave consultations about "how to get over the guilt of having a cash-only practice." There was not a long line. There was also the Energy Enhancement System, a computer monitor that hooks up to one's body and "regenerates life on a cellular level"; a company selling "cutting-edge saliva hormone testing to detect your biological age"; a pavilion selling "colon hydrotherapy stations" that can "complement your business

(No Messy Leaks!! NO Messy Blowouts!! NO ODOR!!)"; a German—surprise—outfit selling "fresh thymus extract"; another offering a full body-immersion unit called Cardio Cor that floods the body with infrared light and hence lets you "ride . . . reduce . . . rejuvenate"; a BioBanc system that lets you store your own white blood cells so that you can be prepared for tomorrow's medical miracles; a supplement called "Pee-Nuts," for "prostate health in a bottle"; a product called H4O, water with hydrogen gas dissolved in it for reducing bullshit oxidative damage; and, my favorite, a natural herbal lubricant called, perfectly, "Virgin Again."

As entertaining as the floor show is, the main medical-science core of A4M revolves around hormones, and if there is a prince of the contemporary hormone world, it is Thierry Hertoghe, the Billy Graham of Thyroid. I first met Hertoghe at an evening session, where his presentation was billed as "Diagnosing Hormone Deficiencies—A Live Consultation." All day long I had heard about him, and I was encouraged to get there early. ("He is so popular people chase him into the bathroom after he speaks!") Turning up a few minutes late, I found the session in full swing, the ballroom packed with rapt physicians—MDs, DOs, homeopaths, and naturopaths. Up on the stage, microphone in hand, dressed in a pink blazer and beige pants and green tie ("testosterone zhust makes everything so *veevid*!") was Hertoghe, a youngish, handsome Belgian doctor with the flopsy hair of a 1970s French movie star. Next to him, his back turned to the audience, stood a middle-aged Asian man. Hertoghe was using him as an example of how to "read" the clinical signs of hormone deficiency on the human body, rather than rely solely on lab tests.

First, he assessed the man's overall appearance. "I guar-
antee you that if you take HGH, you will definitely stand up
taller," Hertoghe said. He then had the man raise first one,
then the other of his bare feet, which he tapped with his fingers. "I see some discoloration, some yeast there," Hertoghe
said. "Do you eat a lot of sweets?" The man nodded. Hertoghe
went on. "Hmmm, and I see some edema in your buttocks."
There were laughs from the audience. Hertoghe's pointer was
on the man's calves. "Oh, sorry, I mean your *calves*, that's
right. That's DHEA deficiency. Now look at his belly—there
is bloating, he eats too much and too fast. Now, what should
he do?" The audience was still. "One thing is easy. He needs
thyroid. I would also give HGH—that would straighten your
face out. And maybe some testosterone?"

Next up was a sixty-year-old Asian woman named Cindy,
who complained that "I have too many food allergies." Hertoghe immediately pointed out that her hair was rough and
dry—"maybe some thyroid deficiency"—that she had no
eyebrows—"thyroid and possibly growth hormone"—and
that "your face could look firmer—that would get better with
growth hormone." He went on. "Your inside hands have a
brown pigmentation—that's cortisol. Your palms are wet—
cortisol also. And look at your fingers, so bloated! You've been
thyroid deficient since childhood!" He also told the woman
about her own personality. "You tend to get dispirited when
you fail. That's cortisol too." He then turned to the audience
and made his diagnosis: "So, the whole body type is thyroid
deficient since childhood, and then HGH and cortisol. I
would supply them all at low doses. And I would have her cut
out grains and milk." He stopped and looked at Cindy and
shrugged. "I bet zee food is bad too, right? Zee way you eat?"

After the session, I lined up with a number of people who wanted to talk to Hertoghe. One man had him look at his eyes. Another at his upper back. It was an amazing display, almost virtuoso, although I could not tell what kind of virtuoso— legit or not. Was he practicing endocrinology or hormonal mesmerism? In a way, it all felt very Old World, perhaps the way medical men assessed people before the era of blood tests and X-rays and CAT scans and MRIs. I had asked sober Fran Kaufman, my endocrine reality check, about the use of such clinical signs. "We used to call them endocrine pearls," she said in her usual equable fashion. "And they have some basis, but they can be very wrong and misleading without the lab tests. I mean, I saw a friend I had not seen for a while the other evening, and, if I were purely 'reading' his physiognomy, I'd have to say he was insulin resistant, maybe testosterone deficient. But then again, maybe he is just chubby. You don't know without the labs." When I came to the head of the line, I asked Hertoghe about it. "You are right," he said, before conveniently tacking the question in his own direction. "You can have access to all the hormones you want but you've got to have the skill to really look at the body. It is an art. A lost art." Then, as a line of men and women tried to follow Hertoghe to the bathroom, he was gone.

But where and when, I wondered, was "it" lost?

Hertoghe, as it turns out, is more multidimensional than his presentation suggests, and when I began sketching out his biography, I found he had left a memorable legal trail, both in his native Belgium and in the United Kingdom. In 2007, he prevailed in a defamation lawsuit against the Belgian National Medical Board; Hertoghe had been running for an office on the prestigious organization when his ensconced opponents issued a booklet saying, in essence, that all antiaging physicians were quacks. The Belgian courts sided with Hertoghe and, in 2007, condemned the National Medical Board for calumny—a wonderful Old World sin—and falsifying elections. He prevailed in a separate case in the United Kingdom as well, where his testimony won exoneration for a London physician accused of being too free to hand out thyroid. When I finally caught up to him, he was exhausted. Where he'd felt "branded" as a quack before, now he was experiencing *le fugue d'victor.* "In Brussels there has been a lot of almost paranoia about me, like 'Don't say andropause, or Hertoghe will sue you.' But the fact is that I have the stomach to straighten out people who libel antiaging doctors, unlike Goldman and Klatz." He jerked his head, Jean Claude Belmondo style, at one of Klatz's posters. "*They* didn't have the stomach."

I asked him why established medical folks seemed so quick to judge the field of hormone replacement and anti-aging. He acknowledged the usual: the up-and-down history of hormones, gerontology, and nontraditional medicine—yes, it was a problem. So was nomenclature; it might be better to get rid of the antiaging label and instead use, say, "longevity medicine." But he also detected something else. "Many MDs have a sense of shame—that there is this huge world of hormones that they were never taught about—and it is a shock to hear about how much they don't know. I mean, let's face it, there are 5 to 6 billion people out there who are prematurely aging. They can't do anything about it." He stopped, perhaps to let me digest the fact that, to Hertoghe, everyone in the world is aging too rapidly, then went on to speak about his mission to slow down the aging

process, and how he has had "mystical experiences" related
to aging. He talked a bit about the purely technical medical issues—about how the reference ranges for determining
what constitutes a "normal" versus "treatable" hormone
level should be narrowed, thereby giving physicians more
elbow room in prescribing hormones—but the more Hertoghe talked, the more one subject dominated: his family
history, and the early twentieth century.

Hertoghe hails from a long and illustrious line of Belgian physicians credited with making pivotal contributions to the field
of mainstream endocrinology. His great-grandfather, Eugene
Hertoghe, was one of the first modern medical men to diagnose and treat myxedema—thyroid deficiency—in the late
nineteenth century. Eugene Hertoghe's breakthrough was in
deciphering that people with shortened eyebrows (now known
in medical literature as Hertoghe's sign) or none often had a
serious thyroid deficiency, which could be cured via an oral
thyroid extract. In other conditions, such as cretinism, it could
be prevented or cured with iodine supplements, which we now
get through table salt. In the early twentieth century, Eugene
garnered a sizable American following, often doing live consultations at medical conferences. In 1914, in New York, he
spoke before a huge audience at the annual International Surgical Conference, many of whom were concerned with the
debilitating aftereffects of goiter surgery—the removal of a
chronically swollen thyroid gland. Eugene Hertoghe described
one such case, cured with thyroid extract, in almost magical
terms:

She had the drawling voice, the sluggish attitude of body, the thinned hair and eyebrows, the swollen mucous membranes, and the difficulty in swallowing. Her tongue appeared too large for her mouth, the floor of which was swollen and raised till it suggested a double ranula [cyst]. Even her ocular conjunctiva was edematous and prolapsed while her complexion was amber-yellow with patches of red on the cheeks. She had also the low temperature with subjective sensations of cold—but indeed Mme. X. presented all the symptoms described in post-operative myxedema and if so she must suffer from spontaneous myxedema.

As my last diagnosis was made, M. X. was too polite to say that he did not believe me, though his face plainly showed his incredulity, but he followed my instructions to the letter.

The result exceeded my hopes. After three weeks treatment the bodily and mental transformation was so complete that she would no longer have been recognised as the same woman. The edema of the tongue, of the lips, and of the eyelids disappeared as if by enchantment and the face assumed an intelligent expression.

So did Eugene Hertoghe set the stage for the family destiny: hormonal medicine. His son, Thierry's grandfather, and *his* son, Thierry's father, both followed the elder Hertoghe's course, with Thierry's *pere* credited with discovering and characterizing one of the key adrenal hormones. By the time young Hertoghe—tall, handsome, perhaps a little cocky—began to contemplate a career, the "incredulity" and "enchantment" of it all had worn thin. "I actually never meant to be a doctor because my father was so passionate about it all the time." He instead became a psychiatrist. "But then I thought I was having hallucinations—because everywhere I looked, it seemed I saw hormone deficiencies." What followed was a kind of medico-spiritual odyssey. Hertoghe went to London to meet Katrina Dalton, who had pioneered the diagnosis of premenstrual syndrome and its treatment with progesterone as a counter to estrogen. He took her ideas back home. "At the time, estrogen was only given at menopause, but I saw that there was a cat-a-*stroph*-ic shortage earlier in life," he recalls. "I began giving it to a forty-five-year-old woman, with proges-terone, and I began seeing that the aging in her face was cur-tailed. With others, I began to understand that a woman who's just had a baby looks like a mother—but what she really looks like is an older woman. If you supplement with estrogen, she looks young and happy again. She's a 'sexy lady.'"

Hertoghe also wondered about the other sex hormone—testosterone. He struck up a friendship with Georges Debled, who had been popularizing testosterone therapy in France (and whose illustrations of "le silhouette de l'homme regressif/progressif" inspire the drawings that begin this chapter). "He sat me down and drew these pictures for me, of what hap-pened when both men and women were supplemented with T, and it was a revelation to me." Debled sent him to another pio-neer, Jens Müller, in Denmark, who popularized a theory of cardiovascular disease and hormone insufficiency. "He didn't want to see me at first when I went to see him—I literally had to stick my foot in the door—because he thought I was too good-looking and not what a doctor should look like, but we eventually became friends."

With all of these insights accumulating, and young Thierry's chatter about it in family circles escalating, it wasn't long before Hertoghe's father, who had continued the "live consultations" pioneered by old Eugene, asked the son to fill in for him. "It was completely 'on the spot,' and I had to rely on my learning very quickly," Hertoghe recalls. "But it was clear I had some gift too. I hate to say it that way, but right from the beginning, when a boy came up from the audience for an exam, I could immediately detect specific thyroid deficiencies." He took his presentation to the United States in the early 1990s, where he floundered for an audience, until he met John Grasela, then just getting his San Diego compounding empire off the ground. Grasela recalls that "Hertoghe had maybe fifteen people at a session in Florida I remember, and I recall thinking, 'This guy is great, but he knows nothing about business and marketing himself.' " Grasela took care of that. Hertoghe's audiences exploded in size, evangelist style.

A month or so after our talk, I attended one of Grasela's "transform your practice into an all-cash business" workshops in, where else, Las Vegas. For three days, about 250 physicians, who had paid $950 each, sat and took detailed notes. There were presentations from a number of established antiaging practitioners. Hertoghe, clad in a vivid pink sport coat and beige cargo pants and cream shoes, was the star. Systematically, he went through each hormone and its clinical manifestations—how they show up on the body. "And remember," he said as he began. "You cannot do a good hormone assessment without touching the patient, and I know that can be a problem here in the United States."

One thing was utterly telling: Many of the slides were not

new—they were from great-grandpa Eugene's time. Here was a goiter. Here was a cretin. Why was that? We don't have goiters and cretins now—we have table salt, I wanted to say. What about the great "in between," the untreated low hormone levels that Hertoghe had championed so much, the ones that accelerate the aging process and, if treated properly, can make the ladies sexy again? There were some minor cases, but what soon slid to the surface was Hertoghe's aesthetics. To wit: modest hormone deficiencies being too difficult for most physicians to spot, perhaps it is best simply to show "the ideal body." Everything that departed from that, the assumption went, must be a hormonal deficiency. "We have to see the extremes to see where our patients should be." He paused and looked at us. "So what *is* an example of a hormonally adequate body?" The slide on the screen changed. It was Michelangelo's *David*. "See?" he said. "There's a firm face. Have we all forgotten what a good firm face looks like? If not a firm face, then he's not in good health. The same with the behind. It has to be firm, too." He flashed a photo of a fashion model. "So you have to have them take off their underwear and look." He went on to estrogen deficiencies. Again, more models, although whether he meant them as an ideal or as an example of deficiency was unclear.

Everyone is prematurely aging? Everyone should be on hormones? If one doesn't have the *David*'s "tone," then one is an estrogen case? All of this began to make me feel the way I'd felt with Paul McGlothin. Had Hertoghe played too much soccer with his head? Or had I? Everyone else was busy writing down everything he said; next to me was a Yale-trained MD, scribbling like a Boston lawyer.

Before I could completely digest all of this, Hertoghe moved on to my favorite hormone, testosterone. He showed slides of the key clinical signs: a flat hair line with no temple wings, pale skin, flabby breasts, a fatigued overall countenance. "This is a person who is slowly but certainly being castrated by time." He paused for effect. "Castrated by time!" As if on cue, the guy next to me put his hands over his crotch. Hertoghe pointed to the man on the slide. "This person suffers. Remember, we are not made to age. We are made to be healthy." He went through the signs of growth hormone deficiency, then on to the adrenals.

The adrenals are the next big thing in antiaging hormone replacement, probably because many of its manifestations—browner skin, a hollow face—resemble stereotypical aging. Hertoghe related some of the less obvious signs of cortisone deficiency. "I see them all the time, and I can almost tell you what their personality is going to be like just by looking at the hands. To them, the whole world is unjust, nothing is ever right, everyone else is unfair to them."

To say that this approach has a few scientific problems, let alone some basic issues of sensitivity, is like saying that perhaps the air in Beijing occasionally might need some improvement. But let's say, for the sake of argument, that Hertoghe himself is gifted, that by dint of family history, experience, inclination, and experiment, he really can spot the deficiencies and treat them. How would his observations help novice doctors do the same? I spoke with him again and he clarified: One would still need lab tests, but the clinical signs were a way to sensitize physicians to the hormonal aspect of health, which to Hertoghe is everything. If one dug in and really thought about such signs, "you will never look at yourself the same way in the mirror again," he said. "Instead, you will see your destiny." I then asked him about my destiny, telling him that the thyroid and testosterone had hardly cured my sluggishness. "Then your food is bad," he said. "Cut out all grains and you'll see improvement in twenty-four hours." He was, in short, simply a hot version of McGlothin; he wanted me to treat my body like a test tube.

Just as I was considering never looking in the mirror again, I remembered something. I had brought with me a picture of Alvise Cornaro, a photocopy of the Tintoretto painting that hangs in Florence's Pitti Palace. I showed it to Hertoghe. "Oh, my God, he has *all* the hormonal deficiencies," he said. "The hair, the eyes, that's estrogen deficient. The caved cheeks and that wispy beard—probably adrenal and growth hormone deficient. He was probably not sometimes such a nice fellow."

From all of these discussions, I drew a few reasonable semiconclusions: Growth hormone likely does not expand maximum human life span, but if the antiagers and their foes can ever take their heads out of their vested interests, it may turn out to be a way to ameliorate the misery of aging. Simply "topping off" hormones—replacing "only what one loses naturally"—ultimately turns into a medical money decision, where one replaces reliance on pharmaceuticals with reliance on lab tests and physician "monitoring" fees. And replacing everything early on—à la Hertoghe—turns one into a bottomless test tube, hopefully with a bottomless wallet. It seemed that there were two models in the commercial antiaging game: Either one had a bunch of science and had paired it with a product that really did not perform, or one had a product that performed but for which the science was inconclusive at best. It all began to sound a lot like the traditional pharmaceutical industry's profile for chronic disease products—think Lipitor for kids, which, as of this writing, Pfizer is.

I had met Jonathan Wright, considered by many the elder statesman of hormone replacement in the United States, at John Grasela's hormone cashfest in Las Vegas. I had also heard about him from one sober-minded scientist, James Duke, one of the nation's foremost medicinal botanists. Although he is no friend of traditional pharma, Duke extends his tough-mindedness into the vast realm of natural and alternative medicine as well, and so I was surprised, when I mentioned Wright, when Duke wrote, "Generally speaking, I find that Wright usually is ... right, especially on the subject of bioidentical hormones." You might not agree with all of Wright's social-political commentary—he's an ardent Libertarian with utter disdain for the FDA, which once raided his offices for prescribing vitamin B_6—but Wright has something substantial, if not completely proven, in his arsenal. It is this, as he put it himself: "Natural molecules are better than extraterrestrial molecules."

In his signature Paul Lynde-meets-Marcus Welby style, Wright means that drugs made from botanicals and minerals that molecularly mirror human hormones, rather than, say, mare's urine, generally work better, with fewer side effects and adverse events. These he calls bioidentical hormones, many of which derive from soy and yam; they are not—repeat *not*—products such as the ubiquitous yam and soy creams sold on the Internet, but pharmaceutically processed plant molecules that are recognized, for the most part, by the FDA and the standards-setting U.S. Pharmacopeia board. They can be obtained only by prescription. If you look at a molecular diagram of, say, estrogen made from mare's urine, Premarin (the "prem" in Prempro), and USP estrogen derived from wild yam, and then compare it to progesterone as made by the human ovaries, the USP progesterone diagram will more closely resemble the human hormone; hence, Wright's use of the word "bioidentical." It is not a term recognized by the FDA, and so, from time to time, Wright and his followers clash with the agency.

When I met him at his spacious offices in his Tahoma Clinic in Renton, Washington, Wright had just come off a jag of publicity surrounding a full-page ad he had placed in the *New York Times.* In it he ridiculed the FDA's recent persecution of pharmacies offering bioidenticals for antiaging purposes. The agency had gone after the pharmacies because, in its somewhat tortured logic, the use of the term "bioidentical" somehow implied that something was natural and better, an assertion for which there were no FDA-vetted clinical trials, and therefore one couldn't say it in an advertisement. "We need someone to sponsor a medical free speech bill," Wright told me, arching a silver-white eyebrow. "Pharma has their DTC ads. We need to be able to communicate fair science freely."

The Marcus Welby part of Wright grew from a predictable past: Raised in a small, confining Ohio town, he was a precocious student who graduated from high school two years early and was then recruited by Harvard. He leaped at the chance and majored in cultural anthropology. ("Um, small

town? I wanted to learn about the world out there.") He went on
to the University of Michigan Medical School, where he studied family medicine—not exactly a hot area in 1969, when spe-
cialization dominated. It was then, there, that the ironic side of
Wright's medical personality emerged; things were not all that
they seemed in big medicine, and there might not be much
one could do about it. He wondered why the university kept its
huge collection of works on homeopathy and natural medicine
under lock and key. "They didn't want us to even *see* it. What,
were they afraid of the competition?" Later, as an intern at
Washington University Medical Center, he began treating a
number of women who experienced complications from birth
control pills, many of which he found could be alleviated with
nutritional supplements—B$_6$, C, and E. "It gradually occurred
to me that what I'd been trained to use did not come from
molecules in nature." That sparked an interest in hormonal
deficiencies, and, because of his success with vitamins and
minerals, the possibility of making a product that would more
naturally resemble the human hormone molecule.

Today, flooded with natural products advertising, it is easy
to downplay Wright's innovation, and, because oversell inevitably produces skepticism and cynicism, to dismiss it entirely,
as have many in the leadership of traditional medicine, especially geriatrics. One might quarrel with many of Wright's
other issues, his borderline conspiracy chatter about the FDA
and drug companies, but what Wright did with hormones is
clearly worth a second look. Thousands of elderly Americans
continue to benefit from them today, and if insurance companies could get better data on them, they might find them useful in developing a lower-priced geriatric armamentarium.

Wright commenced work on bioidentical estrogen in the early 1980s. At the time, hormone replacement therapy for women was undergoing one of its perennial reformulations. The principal change grew out of estrogen's implication in endometrial cancer, which had been ferreted out by researchers in the 1970s. The findings were clear: Estrogen is a growth factor, and, not properly balanced with progesterone, another human hormone, it can lead to excess and malignant growth. The pharmaceutical industry responded by repackaging its estrogen, Premarin made from the urine of pregnant horses, with progestin, a progesterone-like molecule also made from mare's urine and branded Provera. Provera—the "pro" in Prempro—was made to play the role of progesterone, the ovary's natural estrogen controller, or "opposer." At all of this Wright scratched his head. Something didn't feel right. He looked closely at the composition of Premarin: Of the three main estrogens, it was composed of 75–80 percent human estrone, 6–15 percent equilin, or horse estrogen, and 5–19 percent human estradiol and other horse estrogens. He then asked: what are the average estrogen proportions in humans? He talked a local blood lab into some records research and found that the average human had three major estrogen molecules in play, in the following percentages: estriol, 60–80 percent; estrone, 10–20 percent; estradiol, 10–20 percent. The natural question: why had the drug companies abandoned estriol? The answer: estriol had always been characterized as a relatively minor, weak estrogen, at best a by-product of estrone and estradiol.

"But that was a wrong assumption," Wright says. "If you look at the best reviews—of six decades of estriol research—

what you find is that scholars are constantly concluding that that makes no sense. One of them said it perfectly: 'It would be unusual if nature produced three estrogens of which only one is utilized.' " More, the natural estrogen supplements, first derived in the 1930s and dismissed by many as too weak to be useful, had undergone a revolution as well, and, through trans-dermal cream and patch technology, were absorbed just as well as Prempro pills. (Wright makes much the same case for testosterone supplementation for men: Use frequent, low-dose transdermal applications rather than infrequent high-dose in-jections to get the benefits without the side effects.)

Wright's innovation, along with those of the proges-terone pioneer William Lee, was to seek out natural but pharmaceutical-grade sources of the three estrogens, and then compound them in human physiologic proportions. These he dubbed bi-est and tri-est. He then used lab tests to determine the rate at which patients were absorbing them, making dose and composition adjustments in response to individual varia-tions. The results, he has demonstrated in hundreds of case re-ports, are remarkable and consistent. Used with compounded natural progesterone, natural hormone replacement—NHR as he has named it—seems to confer most of the benefits of pharmaceutical HRT, with fewer of its side effects. Because his various compounds have not undergone gold standard clin-ical trials ("We just can't afford them," he says), he is forever hounded by the FDA for making unproven statements. The FDA, in turn, is frustrated by something else: Wright may be right. In the agency's own pending review docket are expensive new drug applications from large pharmaceutical companies looking to brand estriol, the once-discarded, "weak" estrogen.

Another frustration for Wright's foes: some research on hormone replacement seems to suggest what Wright has been saying about dosage and manner of application. Low-dose estrogen therapy commenced before or upon menopause may confer the benefits—from better blood fats to better bone and brain health—without the increased risk of cancer and heart attack. A parallel effort reassessing long-term, low-dose testosterone replacement in men is also under way, with a number of studies suggesting that some form of testosterone replacement may delay onset of age-related dementia.

In struggling through all of this, Wright has discovered much, he says, "that has been lost." He has the propensity to extol the "ignored" past, perhaps without fully acknowledging the solid medical science that was responsible for burying that past. Like Hertoghe, his reference points gravitate toward the early twentieth century, the so-called golden age of hormonal medicine. His favorite "lost treasure," and one he cites frequently in speeches and writing, is a man named Henry Harrower. Harrower was the founder of pluriglandular therapy—the notion that human hormonal deficits can be treated with extracts of animal glands—and the president of Harrower Laboratory, which sold said extracts. If nothing else, his story shows just how permeated was Jazz Age America with hormones, even at the small-town level.

Harrower, a big, handsome man with a well-trimmed barbershop mustache and flair for nice clothes, originally hailed from England; after earning a degree in massage therapy in Sweden, he landed at Battle Creek, Michigan, where he earned a medical degree and fell under the spell of John Harvey Kellogg, the Andrew Weil of fin de siècle America. He later traveled back to Europe, where he became friends with Eugene Hertoghe and learned the fundamentals of thyroid therapy. Swimming in semidiscoveries and half-surmise, he confected his own unified theory of hormonal action. Reasoning outward from thyroid's success as an oral supplement, Harrower said that all hormonal deficiencies could be righted by a process he called homostimulation—ingesting the extract, often desiccated, of corresponding animal glands. Desiccated sheep's adrenals could be used for human adrenal failure, extracted cow pituitary for human growth failure, tonics of bovine ovaries for human estrogen deficits, and so forth. He concocted a universal dosing theory, which he dubbed Harrower's theory of hormonal hunger. The essence: one could not give too much glandular extract, because the body knew how to pick out just the right amount and discard the rest. He was a vigorous synthesizer of medical science, and, even after being thrown out of the Endocrine Society, which he cofounded in 1919, for his obvious conflict of interest, his ideas loomed large. (Harrower Laboratory became the largest employer in the Southern California city of Glendale—known in that time as "Gland-dale," and Harrower himself may have been Aldous Huxley's model for one of the protagonists in his anti-aging novel, *After Many a Summer Dies the Swan*.) Through journal articles and his laboratory's monthly brochures and newsletters, Harrower prefigured much of modern-day doctor detailing. As Harvey Cushing observed of it: "Surely nothing will discredit the subject in which we have a common interest so effectively as pseudo-scientific reports which find their way from the medical press into advertising leaflets, where, cleverly interwoven with abstracts from researchers of actual value, the administration of pluri-glandular compounds is promiscuously advocated for a multitude of symptoms, real and fictitious."

It was Cushing, perhaps the greatest of early modern endocrinologists, who destroyed much of Harrower's credibility. This he did in a famous presidential address to the young Endocrine Society in 1921. Aside from thyroid, Cushing said, endocrinology was in a primitive state, "so what is there to say on pluriglandular complex except to acknowledge an abysmal ignorance?" Hormonal hunger was "buncombe." Homostimulation too: "The Lewis Carroll of today would have Alice nibble from a pituitary mushroom in her left hand and a lutein one in her right and presto! she is any height desired." As the medical historian Theodore B. Schwartz acutely notes of Cushing's address, "This presidential diatribe was published not in *Endocrinology*, but in the *Journal of the American Medical Association*, where it would receive the widest attention."

I asked Wright about that. Given Harrower's faulty science of dosage and reliance on oral administration, did he really think that returning to his ways was wise? Wright acknowledged only one problem—that Harrower relied too much on oral dosing, which was ineffective. Whole gland extracts, administered intramuscularly, were "absolutely the way to go in the future." I rubbed my liver nervously as he described to me his latest putterings—an experiment with his mother-in-law and injections of the adrenal cells of a fetal pig. "We did this for three to four rounds, and all of her friends wanted to know, you know, 'what spa have you been to?'" He had a study from Europe—from which century he did not specify; it showed that nearly 75 percent of tissue cell injections reach their target organs. To his mind, that suggested that all kinds of cell and gland therapy will soon become huge in the antiaging field. "Not just stem cells, but all kinds of cells." Glaucoma can be alleviated with shots of lamb's cortin, an adrenal extract; the same for hearing loss. Both were heavily promoted for such in the 1930s, before the hormone receptor theory came to dominate.

For $75,000, in fact, one can go to the offices of David Steenblock, an Orange County osteopath, and get "autologous bone marrow stem cell therapy" for just about any medical problem. I first met Steenblock at a conference on aging in England, where he was to give his presentation to a crowded Cambridge University conference room filled with antiaging activists, mainstream scientists, medical entrepreneurs, and assorted wingnuts. He is a convivial man, balding and routinely dressed in Tommy Bahamas Hawaiian shirts and tan slacks and loafers. Despite that, I did not hate Steenblock. Over breakfast, I caught him casually bad-mouthing the previous night's presentation. "Oh my God, could you *be* more complicated?" he said. His own slide show, a few hours later, was not. Pointing to slides on a jumbo screen, Steenblock walked the audience though his typical procedure, one laced with references to science. As he explains it, he first places the patient in a hyperbaric chamber, the better to increase stem cell production in their bones. (There is some evidence that this works.) "Then we, kind of, er, hurt them a little," he said, apparently to stimulate more cell proliferation. He did not clarify what he meant by "hurt a little," but I assumed he likely bruises the patients in the shins rather than, say, kicking them in the testicles. Steenblock then uses a needle to withdraw marrow from one of the patient's larger bones, puts the fluid through a kind of purifying process, and then reinjects the fluid back into the part of the body that needs repair. "And, let me add, this is entirely legal, and doesn't need FDA approval, because you are just using the patient's own cells—you haven't added anything." Steenblock claims a wide range of improvements from the procedure. Nevertheless, he has been investigated regularly by the State of California and has been disciplined by the state osteopathic board for a number of infractions.

Once the door of his hyperbaric chamber (which had been custom designed and installed for Steenblock) blew out while a patient was in the chamber, causing injuries to people outside the chamber.

He would not consent to an interview, but he did sell me a copy of his latest book, *Umbilical Cord Stem Cell Therapy*.

The longer I mingled with antiaging types, both those with traditional degrees and those without, the more I came to appreciate the reason that Big G hates them. It is this: They are surfing on—and making money from—the establishment's science, long before scientists—and most of the time, even ordinary folk—would deem the science sufficient. And unlike Harvey Cushing, Big G has little faith that such pseudoscience will debunk itself. They may be right. If anything, the marketing is slicker and ever more soaked in science. A whole new category of public relations, in fact, specializes in antiaging science.

The latest trend in antiaging marketing is to hook your product to an established "big" theory of aging, just as pharmaceutical companies hook the selling of Lipitor to a theory of cholesterol and heart disease. At A4M, Noel Patton, a tall and serious-minded former farm implement salesman, and his wife, a former beauty queen and TV weather anchor, were selling something called TA-65. It was a natural product, synthesized from the Chinese milkvetch plant, or *astralagus*. Cost: $25,000. Patton's claim was that TA-65 would slow down the age-associated shortening of telomeres—the chromosome endpieces that Leonard Hayflick insists are a kind of internal cellular clock. TA-65 works by stimulating production of telomerase, a telomere-lengthening enzyme. Patton had obtained the "nutriceutical," or nondrug license to TA-65, from Geron Corp., a late 1990s start-up that received a huge amount of press for its telomerase efforts until it was shown that telomerase might cause cancer; after all, telomere shortening might be the body's way to check uncontrolled cell division. I asked Patton if that bothered him. No, he said. "We are constantly checking the blood work on our fifteen clients and haven't seen that at all. If anything, we've seen a slowing of telomere length loss." That telomere length and aging are, at best, an association, and not a causal relationship, did not seem to bother him either. I asked Professor Rita Effros, an esteemed UCLA pathologist whose work on telomeres and the immune system was used by Geron, if she knew Patton; after all, he was featuring her and her work on his website, and in various lectures he'd chatted about her as if they'd just gone dancing together. She had never heard of him. "Who is he?"

Another way to sell a product based on a theory of aging is simply to take the tools that bench scientists use to measure aging in mice and sell it to people. As Vincent Giampapa, a plastic surgeon and cofounder of A4M tells it, he was "looking for a way to affect human aging on a deeper level" when he began reading about the free radical theory of aging, and the ways in which mammalian scientists, like Masoro and Walford, measured oxidative stress. He then began using it on people, determining first their baseline level of DNA oxidation, then taking blood tests after his patient/subjects had been on an extract of the herb known as cat's claw, or *Uncaria tomentosa*. He saw that certain markers in a group of fifteen patients declined after administration of cat's claw for four months, and then began selling the tests ($285.00 apiece) and the herb ($69.95), along with a book, *The Gene Makeover*. Although a trial of fifteen people for four months would never pass muster at the FDA, there remain even deeper problems with Giampapa's approach. "These tests are, at best, preliminary and tentative, even in the world of rodent science," says Arlan Richardson of the Barshop Institute at the University of Texas Health Science Center in San Antonio. Richardson has been studying oxidative stress, aging, and mice for thirty years, and one emerging aspect of that work is deeply troubling, not only for people like Giampapa, but anyone who claims to base an antiaging cure on the free radical theory of aging. "Recent findings do not bode well for the free radical theory," says Richardson. "We are seeing mammals that live very long lives, but with lots of free radical damage. We are seeing that oxidative stress may be important to specific disease, but not so important to the underlying aging process. Oxidative stress may simply be a gray hair."

But gray hairs matter, and, in the world of commercial antiaging medicine, so does one other thing: skin. Treatments to ameliorate its deterioration, from wrinkle cream to Botox, make for a $50-billion-a-year industry alone—and that's just for the stuff that doesn't work, stuff that does not affect the basic aging processes. In fact, the basic aging process of skin seems so intractable as to preclude any serious discussion of altering it. Consider the basics, as laid out by none other than Edward Masoro, the rodent aging scientist who came to know so much about fundamental human aging at the cell and molecular levels. In his classic work, *Problems of Biological Aging*, Masoro covered the basics of skin aging. As humans age, a number of changes begin to occur at all three levels of the skin. In the upper, most visible level, is the epidermis. In it, three things go wrong: It stops shedding old cells, it slows down the process of replacing them, and it stops making new cells that fight infection and disease. At the middle layer, the dermis, the process gets even weaker. The two things that make skin youthful—the proteins known as collagen and elastin—begin to shrink and dry up; the surrounding area, the so-called matrix, gets stiff, while blood vessels grow fragile, encrust, and break. The third, or subcutaneous fat layer, grows thinner. All of this causes, as Masoro might put it, very unsuccessful aging. You can play around with hormones, and you may well see some skin improvements in that process, but what, outside of hormones, can slow down that internal process, or even, as the A4M crowd might have it, "reverse age" it?

I first noticed the work of John Shieh, MD, on a website called "aging backwards." It was the typical bonkers net magazine, but it led me to Shieh's own site, and eventually to his practice, in South Pasadena. There, he has opened a clinic called RejuvaYou, where he specializes in using a combination of low-level laser, radio waves, and heating elements to, as he puts it, "trick your skin into making more collagen." A congenial forty-year-old graduate of a major U.S. medical school,

Shieh had come to antiaging medicine via an increasingly familiar path. He'd got sick of the way contemporary medicine
was being practiced. "I was seeing twenty-five patients a day
in a family practice, which is what I had always wanted to
practice, but I started seeing too many unanswered issues in
patient care, like, what happened when you put people on
statins and they have muscle weakness, what else did modern
medicine have to offer those people? Nothing. I had gone to a
few A4M conferences, and I saw that had something to offer.
Of course, my partners were a little put off—'Why is John
ordering testosterone tests for a thirty-five-year-old?' And I'd
have to explain to them something that seemed self-evident
to me. That, here you have some guy who's doing all the right
things, eating well and exercising, and here his buddies are all
getting muscular and he's not. Why not treat somebody on the
so-called low-normal end if it is affecting his quality of life?"

He began using some of the A4M arsenal on himself. He
tried a growth hormone stimulator. "It was amazing—not
fast—but remarkable. I remember sitting up one evening
and feeling, gosh, I have all of these ideas for how I can make
my practice better. It was like being twenty-five years old
again! I can conquer the world. Or I would look in the mirror
and I could see, hmm, I'm a little tighter here, I can see my abs
here . . ."

Then he got the chance to try out a new skin rejuvenation
system, one known as ELOS. The system was created by an
Israeli company called Syneron and was an attempt to deal
with a basic problem in all "wrinkle reduction" systems: They
often use such a high level of laser and heat energy that they
end up damaging the skin more than helping it. The idea be-

hind ELOS was that, instead of using high levels of one kind of energy, what if you combined traditional heating and cooling methods with radio frequencies and low-level light energy? A number of studies—almost all carried out with Syneron money—showed that you would reduce wrinkles and make skin look younger without the traditional laser resurfacing problems. The theory trades on an older idea of inducing repair by making cells "think" they are injured; an entire body of theoretical molecular biology, in fact, has sprung up around the arena of "heat shock proteins" that do just that, but it is a nascent field. ELOS simply made a machine to induce them, at least theoretically. Shieh tried it out on himself. "There was no doubt, I looked younger. Ask my brother, who's younger than I am! I look younger than he does!" He tried it out on his father, who'd survived years of cancer treatments, with results so striking that he posted the photos on his website.

But where was the science? None of the studies really proved that ELOS worked by "tricking" the body into making new collagen—the essence of any real antiaging claim. And even if it does, it is a temporary fix; relying on ELOS would mean a lifetime of $500-a-pop treatments, maybe three times a year. What about that? How far, really, do any of these treatments take us with regard to reversing the aging process? Shieh and I both went over Ed Masoro's brilliantly simple exposition of skin aging. "You know, he's right, in that a lot of research still needs to be done on this," Shieh said. "But what you've got to keep in mind is that this is safe, and that people get visible results, and that, in itself, can be transforming." I asked Shieh if he simply meant that if you feel good about how you look, then you are likely to take better care of yourself—

the classic mantra of the modern plastic surgeon. And, by the way, wouldn't society be better off if guys like him remained as family practitioners?

He didn't have an answer for that, but he did have a number of case studies of his own patients. As we reviewed them, one by one, he grew visibly more animated. I had seen this happen with other medical people who, after years of doling out statins and telling sour-pussed sick people to eat right and exercise more, started seeing patients who were . . . happier. He began to tell me a story of one patient, a middle-aged Cambodian woman who'd come to him for skin treatment after traditional laser resurfacing had made her skin look even worse.

"She was a very shy woman who'd hide in the back of a local beauty salon, washing hair, because of some damage done to her skin during years of working in the field in Cambodia during the war. She'd had some inappropriate laser resurfacing and my mother-in-law brought her in and said, you've got to help her!" Shieh put her through a series of complimentary RejuvaYou treatments. "And, as I did, something happened. I realized even I had not noticed that this was a beautiful woman. Her eyes. Her cheekbones. Gosh. Wow!" He didn't see the woman for a while and went to visit her at her old beauty parlor job. " 'Oh, she doesn't work here anymore,' they told me. I said, 'Where is she?' And they said, 'Oh, she got a boyfriend, a very rich surgeon. She doesn't have to work anymore.' "

Am I growing tits? I ask this because, for all the high-flown discussions about DNA damage, glycation rates, GHRKO mice, and caloric restriction, tits, male tits, are the real issue with aging men in America. And I don't want them. Yet I read in the scientific literature that while too little testosterone causes hypogonadism and, thus, gynecomastia, so does too much testosterone. I ask Rothenberg about it. He replies, "It could be that the testosterone/estrogen ratio is low and E dominant with hypogonadism. High T can cause it if there is a lot of aromatization [a type of conversion to E] and even though the ratio is OK the absolute amount of E is over a threshold." Upon examination, it is clear that I have what most aging American men have—chubby pecs, not actual glandular growth, or true gynecomastia.

But Rothenberg's message reminds me how much I have come to think of my body as a test tube. It seems to me that antiaging medicine—just like CR—requires an enormous amount of self-absorption to work, a constant monitoring of enzymes and vitamins and hormones and proteins. I wonder whether, in an age of such tremendous self-absorption, including my own, it's the best way to go. Perhaps I am simply being lazy, but part of me asks: Can't I just chuck all this and go in for a tune-up once in a while? Instead of trying to make a perfectly running car (as in CR), or giving the car all kinds of expensive new fuels (as in hormonal, or cash, therapy), can't we reengineer it occasionally to just keep running?

As I was soon to find out, a whole new field of antiaging science and medicine is trying to do just that.

Saints Cosmas and Damian Attaching a New Leg
to a Wounded Soldier (ca. 1495).

BOOK THREE

Engineering

The great choice for this generation is assisted suicide or experimental prolongevist science.

— DAVID GOBEL, PRESIDENT, THE METHUSELAH FOUNDATION

The assumption behind the cold and hot approaches to antiaging is simple. It's this: By understanding the biology of aging, we can either intervene to slow or stop detrimental processes (via CR and CR-like drugs), or we can replenish certain factors that will help us feel, and, in some ways, actually *be* healthy longer (via hormones). But what if we were to say, simply, that we don't care about that, that we don't care why our car is getting rusty, or that it is running out of fuel, but that what really matters is keeping the car running by vigilantly repairing all damage? Like the vintage car buff whose Model T still looks and runs like the day it came off the Ford assembly line, we can view our body's main enemy not as rust (free radicals) or insufficient fuel and oil (declining hormones) but, simply, as damage. Damage to our internal systems—regardless of its cause—will eventually kill us. Our priority should be to eliminate the damage and then replace or rejuvenate the worn-out parts. Period. Yes, free radicals may cause some of that damage, but if we wait until we can harness oxidative stress in just the right way, well, we'll be pushing up daisies. Yes, lack of growth hormone might be bad, but replacing it for too long, who knows? If we can just get a lot better at repair—that might be the key to push out healthy life span, and, perhaps even more, push out maximum life span. This engineering approach—highly experimental, pragmatic, and controversial—constitutes the third wave of antiaging thought. Its practitioners have offered everything from new kidneys built with your own cells to thousand-year life spans. Much of what you believe about it will depend on how you understand your own body and the processes that make it what it is.

Consider an experiment that almost every American, at one time or another, likely declared as incredibly weird, bizarre, unfathomable, or downright unnatural. It concerns an ear, a human ear that was grown on the back of a mouse. From the first announcement of its creation by two Boston scientists in 1992, the mouse ear, inevitably paired with a mental graphic of a mad scientist holding a test tube, has served as a vehicle for every imaginable fear that the public holds of modern science, from genetic engineering (which it wasn't) to infectious disease (ditto). One image it hasn't been paired with is that of an aging man or woman, but that, if you think about it, is utterly appropriate. The body of practical medical science that the mouse ear was meant to symbolize—that of tissue engineering, or the making of new body tissues and organs—

serves the purpose of life and health span extension perfectly. If it wears out, replace it. But unlike knee and hip replacement, which depends on mechanical additions to the body, or organ transplant, which replaces like with like but is highly dependent on organ donations, tissue engineering is about making new organs and tissue from one's own cells, thereby removing the problems of limited supply and immunological rejection. It is a new science, although the idea has been around at least since the Renaissance, when many artists took to painting the memorable scene of Saints Cosmas and Damian attaching a new leg to a wounded soldier.

Because tissue engineering is such a new industry, it is possible to find and speak to its founders, who are not saints, but rather, a small cadre of Boston physicians and scientists. They are Dr. Joseph Vacanti, a Massachusetts General pediatric surgeon, his brother Dr. Charles Vacanti, and their friend Robert Langer at MIT. The Vacantis hailed from a large Nebraska family of ambitious Italian American stock—all of the brothers went into medicine. Joe Vacanti, a handsome, velvet-voiced man, recalls that "there was never a time I remember not knowing that I was going to be a surgeon." After medical school at the University of Nebraska, he earned a surgical internship at Massachusetts General, working with the late Judah Folkman. Folkman made one of the most important medical discoveries of the twentieth century—that of angiogenesis, or how new blood vessels are born. Folkman had put his observations in the service of cancer research. His was a clear and fundamental mission: a tumor tissue—any tissue—could not grow without a blood supply. Hence, if one could find a way to selectively stop or inhibit the growth of blood vessels, one could stop the growth of a cancer. Such was the origin of what we now call anti-angiogenics. Young Vacanti, his mind expanded by what he had learned in Folkman's lab, found that he was even more interested in pediatric surgery as a career. He had an inkling that some version of angiogenics might be valuable there, perhaps in service of organ replacement. Many of his little patients needed new livers, but demand constantly outstripped supply of donated organs. "I watched in agony and [was] completely helpless as several children faded into coma or hemorrhaged to death," he says. "It occurred to me that if we could build liver tissue, we could transplant on demand."

He and Langer began to experiment, eventually coming up with a polymer surface, or "scaffold," upon which to seed living liver cells. When he stepped back and looked at the results under a microscope, he could see that the liver cells were growing in contact with the polymer, without any toxic effects. The next task was to grow a three-dimensional chunk of liver, and that was where Folkman's work on blood vessels came into play. "The research question was in a way the inverse of what Dr. Folkman was concerned with—we had to find a way to *encourage* the growth of blood vessels. But we found we could use so much of what we had learned in his lab." Every gram of human tissue has one billion cells, and every cell must be within five microns (or five 25,000ths of an inch) of a blood vessel. Certain proteins—complicated growth factors—had to be present in the right amount at the right time. How could they engineer such complexity?

What followed was a lengthy period of contemplation— one not unlike Clive McCay's reveries by the trout pond or Denham Harman's ruminations in the Berkeley labs—during the summer of 1986, when, while vacationing on Cape Cod, Vacanti saw the solution. It was waving gently at his feet in the form of seaweed. "The solution was staring me right in the face! Branching seaweed was nature's way of providing a massive surface area for the plant to extract dissolved gas and nutrition from the sea." He called Langer and the latter set to work designing and fabricating microscopic, ball-shaped clusters of branching, biodegradable polymer. They could now seed this with living cells and grow tiny chunks of tissue. Later they used similar principles to craft ear cartilage, seeding it with living cells, and, famously, implanting it on the back of a mouse to act as an incubator. It worked. "All of this we did, really, as engineers," Vacanti says. "We were scientifically skipping steps—we didn't know every single pathway, every single gene, involved in tissue growth. We were acting on some basic principles from our disciplines. Because the point was never just research—it was 'what could this do for a real patient?' " What followed was a spate of innovation. The Vacantis and Langer built structural tissues for urinary repair, skin, cartilage, bone, and blood vessels. In 2000, the Vacantis and Langer figured out how to use silicon wafers and micromachining— inspired in part by the ways of computer chip makers—to produce delicately branched scaffolding materials. They have already done successful transplants, in animals, of partially engineered organs—heart valves in lambs, partial stomachs in rats, kidney structures in pigs. The ability to build a massive chunk of functioning tissue, like a liver, is still, says Vacanti, "a ways off. I can't even speculate on when."

Although the Holy Grail of whole organ generation for humans has remained elusive, the Vacanti-Langer team's numerous partial victories have prompted others in the field to rethink the whole idea of organ replacement. Perhaps the most radical notion is that, in the future, our organs may not look at all like the organs we have now. If our kidney's filtering system goes, we may simply replace it with an organ, built from our own cells, that only performs the organ's waste-filtering functions, leaving the original organ to do things like process potassium. "Most organs are just a bunch of tubes," says Gabor Forgacs, a tissue engineering renegade at the University of Missouri. "It does not have to look like the human kidney. The point is to be able to quickly manufacture a working organ out of the patient's own cells that will do the job that the organ is no longer doing. We have to make these sort of leaps. Because, I mean, we know we can't rely just on donated organs. Look at the huge demand we have even today that we cannot meet. If you want to live forever, we've got to do better."

A compact, tousled-hair Hungarian with a charming Old World edge, Forgacs is a relative latecomer to tissue engineering. His original degree was in physics, but upon landing at the University of Missouri in the late 1990s, Forgacs found himself caught up in that university's huge investment in tissue engineering, using pigs as the principal experimental model. There he met John Critser (no relation to the author), one of the country's foremost porcine embryologists. Critser encouraged Forgacs to push some of the traditional thinking about tissue development. He was a perfect person to do so, Critser says, because Forgacs was not from the traditional farm belt–animal husbandry world, but, rather, from "a theoretical background." I found out what Critser meant when I met up with Forgacs at the National Swine Research Conference in San

Diego in 2008. There he was giving a presentation to his colleagues about his breakthrough idea—"organ printing," the use of cell clusters and a printerlike device to literally build new organs from the ground up. The genesis of the idea, he said, came from the fact that he didn't like the Vacantian insistence on using a collagen scaffold to guide the shaping of a new organ. "I'm not big on scaffolds for a lot of reasons," Forgacs said. "What is the right scaffold for the right cell? What if it is not fully degradable?" Instead, he said, "I wanted to find a way to get the cells to guide themselves into a structure of their own."

To do that, he had to go back to hard biophysical principles, specifically the well-observed phenomena of tissue self-assembly. In self-assembly, clusters of cells in the embryo sort themselves out into inner, middle, and outer tissue walls. "I was *totally* fascinated by this," he told me later, "because, you know, I was educated in brrrrutally theoretical physics—everything had to have a reason, a place in the schema. What was the determining factor?" The answer was surface tension—the amount and distribution of surface tension guided cells to sort themselves into the correct configuration. Get the cells in the right configuration of surface tension, perhaps aided with various growth factors, he reasoned, and "the magic," as he likes to call it, would happen. By using aggregates of cells as the "bio-ink," a gel-like biopaper and a programmed printerlike device, Forgacs succeeded in creating transplantable mouse and pig arteries, in one case building the rudiments of a working mouse heart. The latter was the kind of big, signature piece that is required to attract government funding these days (which he has), but Forgacs doesn't get really excited until he starts talking about the human kidney. Concluding his talk to the Swine Conference, he exclaimed, "I want to build a kidney! It is such a . . . *stup-eed* organ! So simple. What is stopping us?"

After his talk, another scientist collared him in the hotel lobby and pushed a little: "What is so stupid about the kidney?" he asked. "Which cells would you use first? Would you build its filtering function, or the structure that governs blood pressure regulation or what?" Forgacs had a comeback. "Yes, I have a dream, which is to build a kidney. But I would be happy with a simple workable filter. I would even be happy if, at the end, what I have made is a way to rapidly make branchable arteries out of your own cells. I would rest easy knowing I have done that. That would be a good life's work."

Cells remain the key to tissue regeneration—bone marrow cells, muscle tissue cells, liver and heart cells. But the Holy Grail of cells—stem cells, which can become any kind of tissue—were captive for years by strict limits imposed by conservative federal science policy. In the summers of 2006 and 2007, all of that changed, when scientists at two institutions invented their way around the policy. They discovered a way to take cells from the tail of an adult mouse and make them act like embryonic stem cells. Their technique, referred to as "induced pluripotency," uses viruses to infect adult cells with genes that make them into stem cells, which can then be coaxed to become any kind of cell. The implications of their discovery are huge, and while much work remains to replicate it safely in human cells, the basic scientific principle has stood. In 2008, researchers at Harvard University succeeded in using a virus to reprogram mouse pancreas cells that normally only make digestive enzymes with genes that tell them to make insulin. This they did, importantly, in a live animal with diabetes; its diabetes went into remission. Most believe it is only matter of time until human adult cells can be similarly reprogrammed.

When that happens, Dr. Doris Taylor, a molecular biologist at the University of Minnesota, will be ready. Taylor has been studying stem cells for years, and for some time she has believed that aging is, fundamentally, a failure of such cells. "For most of our lives, endogenous—internal—repair of tissue is the norm, because we have stem cells all over our body," she explains. "Yet we do not heal as well at seventy as we do at twenty. That's because the number and quality of stem cells declines with time. Aging is a failure of those remaining stem cells to deal with disease and damage." The insight— a twenty-first-century version of Cornaro and Galen's notion about aging and the depletion of "radical moisture"—drove Taylor from her original specialty, pharmacology, into cell biology, with an emphasis on cell delivery and cell therapy. As she recalls, it is hard to work in that world, with the vast theoretical potential of cell manipulation constantly swirling about one's mind, without also thinking of something practical, concrete. "We began to think, 'can we use these ideas to make a whole organ?' "

To Taylor, the missing ingredient in such a venture was not cells, but the "3-D matrix" upon which to implant cells and build, say, a heart. In a kind of midwestern Frankenstein moment a few years back, she mused upon one obvious source of such matrixes—cadavers. Why not? If you could somehow drain, say, a heart of all the former owner's cells and then implant a series of stem-cell-like heart cells from a living patient, you might get a working organ. She began to tinker with the idea, using, fortunately, rat hearts. First, how to drain, or de-cellularize, the old heart? After a number of failures, she and a lab associate tried infusing one with a fairly commonplace de-tergent solution. What they were left with was a clear, translu-cent scaffold in the shape of the original heart.

They then recellularized the heart matrix with new heart cells from a young rat. Then, instead of building a large medi-eval platform that could be raised to the castle roof by clank-ing gears and pulleys, where the heart would await repeated lightning strikes, they stimulated the heart with a small lab-oratory device, creating artificial circulation and even blood pressure. Now there was a weak pulse. Taylor then implanted the new heart into the abdomen of an unrelated rat. Not only was the new organ not rejected, but a blood supply began to develop as cells from the host began to populate the heart walls and vessels. Much work remains on certain fundamental questions—"Can stem cells be placed anywhere in the body and still produce a heart or a kidney? Or must that stem cell be placed in a certain anatomic position to do so?" But "it doesn't seem unreasonable to me to use human cells and matrixes to meet this huge and growing pressure we have for new organs in transplants of all kinds."

Perhaps what is most provocative about Taylor's work is the challenge it poses to the status quo: What's next? She has, in a sense, solved the cell biology question. Now the question is one of scale, of how to test the idea on more human-size organs. "The pig is perfect," she says, "the scale and the physiology are right, and you can pump the right amount of blood." Taylor's hopes for the pig led me back to John Critser, at the University of Missouri's Swine Research Center. Perhaps more than anyone else in the country, Critser, a longtime embryologist who has studied everything from mice to elephants, sees the big picture when it comes to tissue engineering. He is the classic out-of-the-box thinker in a traditional field; one of his more remarked-upon experiments, considered key in the arena of rare species preservation, was to implant ovarian tissue from an endangered elephant into a lab mouse to produce . . . an elephant oocyte, or egg. Because he is such a strategic thinker to specific ends, I asked him what he thought constituted the biggest barrier to progress for tissue engineering. After all, tissue engineering has commanded several billions in venture capital over the years, and the NIH itself has committed sizable resources as well. Where were the organs? I wanted to buy one. What was holding things up?

A thoughtful, soft-spoken man with an elegant head of silver hair and a neatly trimmed mustache, Critser at first demurred. I understood. The world of animal experimentation is a world fraught with public misunderstanding, extreme emotions, anthropomorphism—"selling the animal short," as one animal lover once defined it—and just plain queasiness. We talked about the intermediary steps that must now happen to translate breakthroughs like Forgacs's and Taylor's into human reality. We also discussed how that process was so fundamentally linked to animal work, particularly xeno-transplantation—say, pig to nonhuman primate. We wondered if the public would ever be ready for that. How shocking, even for Critser, were the photos that appeared a few years ago on the front page of the *New York Times*, the ones that showed Chinese peasants hand-drying pig intestines for use in making the human heart drug heparin! Well, what *about* the Chinese? I asked. Critser raised his silver eyebrows a little and hesitated again. "China will overtake us with xeno-transplantation, I'm convinced," he said finally. "For better or for worse, they seem to be ready to take the risks that go with it, maybe risks we are not ready to take yet."

There are those, however, who are already thinking about how tissue engineering, rebranded as rejuvenation medicine, will come to the assembly line. Chris Mason, a British physician and a leading public intellectual on the subject at the University College London, says, "the real barrier is 'what happens when a therapy is ready?' How do you scale up production to meet the needs of a big patient population?" You can learn a lot about that gap if you look at how resources are apportioned in the few stem cell therapies already out there. Consider corneal adult stem cell therapy, in which cells are harvested from the patient's good eye, incubated on a contact lens, and implanted over the bad eye. It works. But it takes a hundred lab workers, clad in bulky bunny suits and maneuvering slowly over big bioreactors, to come up with enough cells for one thousand patients. That's fine for small age-related diseases, but not for bigger ones, like Parkinson's or diabetes, he says. "You will solve the unemployment problem before you solve the patients' problem if you continue that way." Automation offers one path, with one company in California, Advanced Cell Technologies, taking the lead with a robotized assembly line for collecting stem cells.

All of that may go a long way toward treating specific diseases of aging, but in more than one way tissue engineering resembles the piecemeal approach of modern geriatric medicine.

What about pushing out maximum healthy life span, or even reversing aging, the coveted prize of the modern biogerontologist? What if someone took the engineering approach pioneered by the Vacantis and, instead of applying it to specific organs, applied it to the body as a whole? Could one engineer a human to age slowly at the cellular level, and so not only extend maximum chronological age but also transform healthy middle age from a 30-year span into a 60-, 90-, or even a 120-year span? And by focusing biomedical resources this way, might one eventually engineer a series of interventions that allow one to live well into the 200-, 300-, and 400-year range?

For decades in the field of gerontology, one dared not even suggest such a notion, particularly if one lived within ambulance range of a state mental hospital. There were dogmas. First, aging was thought to be a universal—all organisms aged; they deteriorated over time. Period. Second, outside of CR in mice and rats, there was no evidence that maximum life span in any organism—yeast, fly, worm, rat, mouse, monkey, human—could be successfully expanded. Life expectancy, yes. Maximum life span, no. Third, aging was random and unregulated. How could you a reengineer a process that voluble? And who would do it? Gerontology and geriatrics were filled with pretty satisfied folk who seemed content playing scientist, not engineer. Yet . . . what if the dogma wasn't true? What if aging was not universal? Or random? What if life *span* could be altered?

Beginning in the early 1990s, the dogma began to unravel, at least in the more freethinking realms of the academy. One salvo came from Caleb Finch, an evolutionary gerontologist at the University of Southern California. Although preeminent in his field, Finch—tall and Darwinesque in mien—was an intellectually restless fellow, constantly working across disciplines to come up with deeper ways to understand the phenomenon of aging. A collaborator with everyone from Walford to Hayflick, he had somehow avoided the intellectual ossification of his fellows in Big G and remained engaged with huge, open-ended inquiries, ranging from inflammation to evolutionary neurobiology to hormonal signaling. In the late 1980s, Finch began to wonder out loud—likely out of earshot of Hayflick et al.—if, indeed, age-related deterioration was universal at all. An inveterate naturalist and environmentalist, he began studying reports from marine biologists looking at fish populations and how they aged. One species, the rockfish, jumped out. One of its geni, the rougheye rockfish, could live as long as 140 years, while cousins of the same species lived only 12 years. Dissected, the old fish displayed almost no age-related deterioration. More: there was little if any decline in the rockfish's ability to fight infections or reproduce, and they showed no age-related increase in cancerous lesions. The fish, of course, eventually died—usually from some form of predation—but the revolutionary point was that age-related decline was not inevitable. Finch dubbed the phenomenon "negligible senescence." Later, writing in the *Annals of Gerontology Biological Sciences*, he expanded the idea: "The prospects for continued increase in human life expectancy are of course unknown, but examples from the natural world suggest that *no firm limit is built into the human genome.* The efforts to modify human aging via drugs, diet, and lifestyle interventions are entirely consistent with the observed plasticity in life histories in numerous other species."

No firm limit is built into the human genome. If senescence—

time-linked deterioration in a species—was not fixed by a
hard genetic program, if it is not the inevitable cause of most
death, could it be slowed down by the human hand? Michael
Rose, an evolutionary biologist at the University of California,
Irvine, set out to explore that question by breeding long-lived
fruit flies. This he did, as he puts it, by "tricking evolution":
He forced the flies to wait until they were older to reproduce.
"The ones who [eventually] do [breed] are those that have already proven they can live that long and have the physiological
wherewithal to reproduce," he explained. "Multiple generations of this procedure makes them live better than twice as
long." The significance of that, he says, is that "aging is in no
sense any basic feature of cell biochemistry." The rebellion
against Hayflickism was now in full tilt. Asked by *Discover*
magazine what his work meant for humans, Rose—inclined
to sarcasm—let loose with a full gun: "There are all kinds of
people who are opposed to us doing anything [about aging].
The Federal Government has this need for us to die on our due
date, so you don't bankrupt Social Security or Medicare. And I
have on a number of occasions heard people give very moving
addresses as to why we should die as soon as possible. I think
the phrase that most stuck in my mind was 'So that we can
know God's love sooner.' And let me just say for the record, I
am all for those people dying. They can go ahead. I just know
other people who don't want to die, and least of all by the hor-
rible and unattractive process of aging, and I don't see any reason why they shouldn't be allowed to go on living."

If deterioration isn't universal, and aging doesn't constitute a basic feature of cell biochemistry, what, then, controls
it? Beginning in the late 1980s and early 1990s, a number of

scientists, inspired by the growing influence of evolutionary biology and the patterns it revealed, took to working with simple organisms—yeast, the earthworm, the fruit fly—to see if there were links between genes and aging. Genes, after all, tell a body which proteins to make, and because yeast and earthworms were fairly easy to genetically manipulate, one could readily mutate specific genes and then measure life span consequences. In 1991, Cynthia Kenyon, a young evolutionary biologist at UC Berkeley, exploded on the scene by showing how two insulin receptor genes in the earthworm, daf 2 and daf 16, doubled the life span of the worm when those genes were mutated. Moreover, the mutations also doubled the health span of the worm. They were Cornaro worms. As Kenyon liked to say, sometimes in the presence of the press, "These animals are magical: they are like 90-year-olds who look and act 45." More important, she broke the old dogma that aging was haphazard and random. If you knew enough about what controlled it, well, perhaps you could control aging. At MIT, Leonard Guarente, and, later, David Sinclair, following their own leads in yeast, worms, and mice, came to similar conclusions. They also focused their search on the insulin pathway and genes they called SIRT. SIRT, like Daf, and like CR, seemed to activate the ancient starvation response, pushing the organism into repair and maintenance mode.

The excitement in gerontology circles was infectious. By 2000, mainstream scientists were saying things—out loud— that only a few years before might have gotten them permanently assigned to teach freshman biology. In one notable *Nature* article, Kenyon and Guarente proclaimed, "The field of aging research has been completely transformed in the past decade. . . . When single genes are changed, animals that should be old stay young. In humans, these mutants would be analogous to a ninety-year-old who looks and feels forty-five. *On this basis we begin to think of aging as a disease that can be cured,* or at least postponed. . . . The field of aging is beginning to explode, because so many are so excited about the prospect of searching for—and finding—the causes of aging, and maybe even the fountain of youth itself."

There was an almost overwhelming common theme, as well. More and more, control of a body's energy use was directly linked to life span extension. Some called it energetics. Others call it "nutrient partitioning." In all this one could hear echoes of Masoro, Walford, and even, when you thought about it, Cornaro. Twenty-first-century science had come all the way back to Cornaro. But would there be a Cornaro of antiaging engineering?

I had heard about Aubrey de Grey some time before I met him, briefly, at a CR conference in fun city, Tucson, Arizona. At the time, he was hanging around with Michael Rae and April Smith, and I assumed that he was a CR practitioner. He is lean to the extreme, and there was all about him the air of Alternative Man—long hair, a hermit beard, a blousy shirt. I knew that he was a Cambridge cell biologist who studied aging, and that he had appeared on numerous TV shows, where he had made a series of outrageous statements to the effect that humans could be made to live a thousand years. In gerontology circles, which I was beginning to frequent, everyone had an

opinion about him. One opinion: he was an irresponsible op-
portunist, a fanatic, a fabulist.

De Grey had managed to acquire this reputation by pur-
posely inflaming Big G. Perhaps the most offended was the
CR-mouse science alliance, the modern-day version of Walford and Masoro represented by the University of Michigan's
Richard Miller, arguably the world's leading thinker on mouse
CR and aging, and his frequent collaborator David Harrison,
a specialist in mouse stem cells at the Jackson Laboratory in
Maine. Miller and other members of the American mouse mafia had sought a large grant from the National Institute on Aging (NIA) to test possible antiaging compounds in mice that
mimic CR's effects, and de Grey, no fan of CR, had made the
mistake of publicly suggesting that, in his opinion, the most a
human could expect from CR was two to three years, so why
bother? In one famous remark, he noted that "the long-lived
mammals that Miller describes are much smaller than normal
members of the same species—not something most people
would impose on their children even if long life resulted." He
even went so far as to compare the mouse mafia to the A4M.
The NIA continued the mouse grant anyway, but the acrimony
remained. When I spoke with Miller, a bearish Santa Claus of
a man, four years after his original exchange with de Grey, he
immediately brought the subject up, unprovoked. "You don't
really want to talk about my mice," he said. "You want to talk
about Aubrey de Grey—that's all you journalists ever want
to talk about! He's ... he's ... *like catnip to you!*" Harrison re-
acted almost identically, all the way down to the catnip—an
odd metaphor for mouse scientists, but, as Einstein once probably remarked of creamed spinach, "Whatever."

There was more to de Grey's list of sins than mere infringe-
ments on professional turf and money. Over a period of about
ten years, he pushed and packaged the tentative notions developing in mainstream laboratories and conference rooms—that
life span is not fixed, that it might be engineered in mammals,
and that senescence can be negligible—and then used them
as a basis to argue against the standard way of moving such
observations into medical science. If we know such things are
true, de Grey would say, then we must stop treating aging as
some interesting phenomenon that we can do nothing about.
Aging is a disease. "It kills fucking 100,000 people worldwide
every day, and I want to stop it," he told a rapt audience at one
of the popular TED intellectual forums. The way to do so, he
said, involved nothing short of abandoning the old ways—of
trying to find a natural, evolutionary pathway and then find
a way to tweak it so that the organism would not deteriorate.
Rather, the task should be either to eliminate the damage—
be it plaques in arteries or amyloid proteins in the brain—or
to make the damage benign. The way to do that was through
engineering. This he dubbed, in a tip of the hat to Finch, "Stra-
tegically Engineered Negligible Senescence," or SENS.

De Grey then went on to market every single theory of
aging as an engineering theory. Consider free radicals. As
de Grey saw it, the problem with that theory was not just
that free radical damage to the cell might be a minor factor
in aging, as Arlan Richardson had been saying, but that the
most important damage done was to the mitochondria's own
DNA. In fact, he said, mitochondrial mutations were the engine of aging not because they hurt the cell, but because they
turned the cell itself into a kind of traveling perfect storm,

one that created huge flows of damaging free radicals, body-wide dysfunctional cell metabolism, and, ultimately, cell death. Instead of trying to stop free radical production, one instead should find a way to head off the whole process by reengineering the entire human cell. That could be done, he argued, by relocating all of the mitochondria's DNA into the protective cell nucleus, where it would be much less likely to get fried and mutated. With a big enough investment, this, he argued, could be done through gene therapy techniques. If it worked, it would be transformative. If one was fond of describing the free radical theory as the rusting of the body, then, with reengineering, you would end up with a body that didn't rust.

To say that the reaction in Big G was intense is like saying that the weather in Burma is sometimes warm. In a major quasi-consensus statement, some of modern gerontology's biggest hitters pronounced the SENS agenda a "farrago" and a "fantasy." De Grey, they went on to say, was guilty of "clever marketing" that allowed him to "short-circuit the traditional scientific channels of new ideas." Personal attacks followed. In an article for *Technology Review*, the physician-writer Sherwin B. Nuland noted that de Grey drank a lot of beer and ate a lot of sweets, and that his wife (twenty years his senior) "lacks a full set of teeth" and smokes a lot of cigarettes. In the same issue, Jason Pontin, the editor, called de Grey a "troll," noting that: "He dresses like a shabby graduate student and affects a Rip Van Winkle's beard; he has no children; he has few interests outside the science of biogerontology; he drinks too much beer."

Beer consumption—the new gauge of intellectual integrity?

There were those in Big G, however, who were not so quick to decry. Instructively, these were the same evolutionary biologists whose pioneering work, now conventional wisdom, fueled so much of de Grey's repackaging of gerontology. Caleb Finch, whose 1990 paper on the rockfish originated the whole notion of negligible senescence, was now one of the world's leading gerontology scholars and the author of the definitive *Biology of Human Longevity*. I asked him why he refused to join his brethren in their anti-SENS effort. "One reason," he told me. "It is totally unproductive. Who is he hurting? I mean, look, Aubrey is a polemicist and monomaniac immortalist. That being said, he has stimulated a lot of new thought on extending the health span that could conceivably allow for unprecedented longevity, as well as deeper understanding of the mysteries of aging. What is wrong with that?"

One wet summer day in 2007, I took an express train out of London to see de Grey. He met me at the Cambridge station, looking a little bleary-eyed from a recent spate of travel. He had come via a very old bicycle, which he pushed as we walked up the cobbled streets to the Eagle Pub, the famous tavern where Watson and Crick first conjured the double helix. I asked de Grey if he wanted lunch, and when he said no, I asked if he practiced CR. "God no!" he said. "Why would I do that? I'll just have a pint."

Like all of today's PowerPoint visionaries, de Grey has a diagram that he insists you consider. He drew it on a napkin for me as I snacked on fish and chips and he drank lots of beer. It looks like this:

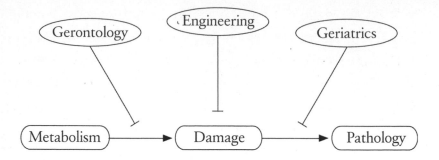

Its simple lines depict what he sees as the basic architecture of modern aging research. The bottom line represents the key aging process—human metabolism. This constant burning of nutrients by the cell and its often poor processing of by-products causes cell damage. Leave that damage to accumulate and you get pathology, disease, and death. Fair enough, so far.

The top line shows where the main branches of aging sciences tend to focus—gerontologists study the processes in metabolism that lead to damage, and geriatricians treat the disease consequence of that damage. In the middle of that line sits de Grey's conceptual time bomb: the as-yet-unrealized role for "a pure engineering approach," as he likes to call it. "You see," he said, "this is all about a repair and maintenance effort. It's not about trying to understand all the possible variables of human metabolism. We don't need to do that to live a lot longer. We just have to identify the key damage and . . . eliminate it."

The analogy that comes closest, he says, is that of the perfectly preserved vintage car. It wasn't designed to run for a hundred years, but it does so because of careful and constant repair and maintenance, using the best tools and materials available. The same with the human body. Focus on the rust that gathers on our fuel lines—the plaque that gathers on and around our arteries—and get rid of it. Forget about interfering or altering the basic process that leads the body to produce it in the first place, the fixation of modern pharmacological science.

To advance this paradigm buster, de Grey has identified seven domains of "cellular interventions" that, given the right scientific and economic support, will stop and even reverse the aging process in human beings. They all bear the imprint of his original thesis about mitochondrial DNA damage—the damage is the thing. His logical tack in describing such interventions, while often zigzaggy, sails a fairly consistent overall course. First, identify a form of cellular breakdown. Start with, say, lysosomal dysfunction, in which a cell's waste-burning organelle, or component, becomes overwhelmed and unable to do its job. That job, as he describes it, is burning up lipofuscin, a nearly indissoluble after-product of metabolism. Now look for the wide-ranging disease possibilities inherent in that breakdown. In the case of lysosome dysfunction, he says, this can range from atherosclerosis (because the cells that attack inflamed arterial plaque can't process the waste and instead rupture and blow up), to macular degeneration (because of lipofuscin-like buildup behind the lens), to Alzheimer's (wherein cells can't keep up the policing of errant plaques and proteins that build up and impair neuronal health). Then ask: what would that lysosome need to be able to break down all that damaging waste? The answer, he says, would be more of the enzyme that it usually uses to do so, but has now been depleted. Find a new source for that enzyme and reintroduce it into the cell. Voilà! Clean veins and clear vision in the year 2500. "The goal is to wipe out the damage using any benign weapon available," he says, "because it's the damage that causes the disease."

What about cancer, a disease of uncontrolled cell growth? Answer: use nanotechnology to design molecular "Swiss Army knives" to "unscrew" cancer-cell surface barriers to oncology drugs. Better still: delete the gene for telomerase, the enzyme that enables cellular division in the first place. How about diabetes-producing visceral fat cells? Answer: stimulate the immune system selectively to target and kill those cells. What about muscle, skin, and organ tissue loss? Replace them with stem cells that reproduce the lost tissue. Amyloid plaque production associated with Alzheimer's? A vaccine that clears the plaque. And how about hardened arteries and weakened ligaments caused by so-called cross-linking extracellular proteins? Answer: after discovering a hyphenation reducer, inject safe chemical agents that break apart the disease-causing links or use nanotechnology to design targeted "molecular buzzsaws."

There are vast problems with all of these—in a sense, SENS *is* a farrago. But what seems to irk de Grey's peers the most is not his approach, but his optimism. Where most scientists err on the side of assuming something will not work, de Grey believes we simply have to assume a 50 percent chance of success. His optimism can take over, sometimes to comic effect. Writing in his 2008 book, *Ending Aging*, about the side effects of deleting the gene for telomerase—

his cancer cure—de Grey says: "One potential side effect of the loss of telomerase from our cells might be eventual sterility for men. If having children is still a priority in a post-rejuvenation world, then men may choose to freeze their sperm in advance."

There is another huge difference between de Grey and his peers. Unlike other pioneers in aging, de Grey's intellectual evolution did not include any level of enrapture with a member of the animal kingdom. There was no McCayian brook trout, no Masoroian rat, no Walfordian mouse. Yes, he knows all of those "models," as he calls them, as well as yeast, worms, and flies. But he displays little sense of wonder at it all. As de Grey tells it, "I've never been able to *not* see through to the end result, and make a decision based on that." He offered up an example. "When I was young, my mother insisted that I take piano lessons, but I soon got to the point where I started asking myself, you know, 'what's the point of this. To be a pianist?' No! I wanted to do something bigger, to make a difference." He was eight at the time.

He progressed through the elite Harrow, and then read computer science at Cambridge, where he was recruited by the software entrepreneur Clive Sinclair to work on artificial intelligence, or AI. De Grey proved himself a brilliant technological problem solver. Computer code—that was his mouse, fly, and brook trout, and he reveled in it. As Aaron Turner, de Grey's programming partner from those days, recalls, "One fateful day as we had been alternately discussing matters both theoretical and practical while perusing the many 'formal methods' books I'd accumulated, Aubrey suddenly slapped the book he'd been reading closed and proclaimed, 'I know how to do it!' From all the pieces of the formal verification jigsaw puzzle assembled in his head, he'd been able to construct a path from problem to solution, which he could 'see' in his mind's eye." De Grey's AI days came to an abrupt end when that ever-nascent enterprise tanked in the early 1990s. By then he had made contact with a number of scientists who needed computing power and expertise. Through his wife, the fruit fly scientist Adelaide Carpenter, he met the evolutionary biologist Michael Rose, who was in the midst of his life span extension work on fruit flies. Rose hired de Grey to help compile a computerized fruit fly gene index, and it was while immersed on the screen that he acquired the aging bug. What amazed, and later infuriated, him, about aging was "that no one was *doing* anything about it!" Why was that? "It was considered a career killer, and it was considered boring. I couldn't believe it. Here was something that kills a hundred thousand people a day and it's boring?" But it was also because of gerontology's strange coming of age in the 1970s. Back then Robert Butler, the first head of the NIA, was fond of saying that calling aging a disease was tantamount to saying that all aged people were diseased. It was like blaming the victim. There also was the ascendency of Hayflick, who, as de Grey saw it, was constantly conflating his in vitro results with in vivo conclusions. Worse, to de Grey's mind, were Hayflick's endless tirades about how life span extension would be a bad, immoral thing. "The result, as far as I saw it, is that a lot of gerontologists who are in their seventies now were sort of stillborn intellectually and ground down by the Hayflick doctrine."

By the mid-1990s, he was in full tilt rebellion against that doctrine, reviving the mitochondrial theory of aging while, at the same time, becoming a transhumanist, the movement founded by Ray Kurzweil that advocated a very de Greyian–style belief: that humans can and should seek a union with machines and technology to produce better, superior humans. Why not? De Grey ordered another pint and went on to explain how Kurzweil's futurism had impacted him and a whole generation of students by "improving our appreciation of likely timeframes."

As if on cue, two fresh-faced undergraduates came up to our table and insisted on talking to de Grey. "Er, sorry, Dr. de Grey?" one of them said. "I just wanted to say that I heard your lecture about antiaging medicine and that I thought it was brill—"

"Then what are you doing about it?" de Grey replied, a little testily.

"What?" his student admirer said, clearly puzzled.

"What are you doing about it?" de Grey repeated, tapping his knuckles on the table before him. "Look, go to my website at Mprize.com and look under 'what you can do.' There's a list of six things you can do to help. Then if you have any more questions we can talk. All right?"

They scooted off, and naughty Dr. de Grey had another beer.

If you wanted to rejuvenate the brain using the engineering approach—focusing, say, on regrowing dead areas caused by stroke and other traumas—you would be hard-pressed to find a more impassioned pioneer than Rutledge Ellis-Behnke. An engaging, balding man with a cherubic countenance and a fondness for medical history—he often cites a three-thousand-year-old Egyptian papyrus on trauma treatment as an inspiration—Ellis-Behnke works as a researcher in brain surgery and neural regeneration at MIT and the University of Hong Kong. As he tells it, the latter provided him a firsthand look at something most academics never see: an epidemic of brain injuries. In China these were caused largely by the sky-rocketing use of automobiles, and, later, by soaring rates of new building construction. "These are lifelong injuries, enormous trauma," he says. "I mean, the number one reason for brain trauma in China for years was bus vs. pedestrian. Think of that. Bus. Pedestrian. That's huge!"

When he would go home to Boston and MIT, he and Professor Gerald Schneider, a specialist in brain and cognitive science, began to tinker with various ways of regenerating neural connections. They faced huge obstacles. For one, if you introduce a foreign material to stimulate growth, you can cause a catastrophic immune response. "And remember, when you lose brain regions, it is not enough to simply replace cells," Ellis-Behnke says. "The brain is like a nursery. You have to find a way to let it regrow the many different types of cells that it needs to do an extremely complex job."

One of the materials he and Schneider played with was a liquid peptide, or biological building block, dubbed RADA16. RADA contained a series of amino acids sandwiched together using nanotechnology—very small scale molecular engineering. RADA was useful because of its unique structure. Its two most hydrophobic, or rejection-prone, amino acids were sandwiched in between the two most hydrophilic, or least rejection-prone, amino acids. Once applied to, say, a cut or wound, RADA did what all good nanoparticles do: it self-assembled. To test its medical utility, Behnke and Schneider turned to the ever-unlucky laboratory hamster. Cutting the key visual nerve systems in several of the animals, they then applied a solution of RADA, and waited. Within twenty-four hours, all of the treated hamsters displayed signs of healing, and within six weeks, vision was functionally restored to all of the adult animals, with more than 80 percent of the severed nerve tracts reconstituted. There was no immune response, no rejection.

What had happened? As Ellis-Behnke details it, the nanoparticles simply self-assembled into a kind of meshlike scaffold, one that perfectly mimics the body's natural neuro-knitting structures, and then, once new nerves have grown, dissolves into a substance that can nourish the newly knit and highly alliterative network. In a sense, RADA does what Vacanti, Forgacs, and Taylor all want to do, but from the ground up. "It is creating a permissive environment in which the body's repair and growth mechanisms can flourish, but without adding new cells itself." RADA, a breed of nanoparticles known as SAPNS, for self-assembling peptide nanofiber scaffolds, has also been tested in living rodent liver tissue, with similar success. Human application might come as soon as three years; faster, of course, with more money. But as Ellis-Behnke sees it, nano materials bring with them something else for future consumers of antiaging medicine. It is a different way of thinking about the body, he says, and, as he often does, he turns his thoughts to history. "With the pyramids, the Egyptians were building things beyond their comprehension with materials they didn't understand. We are at the opposite end. We are building extremely small structures where we do not understand the normal rules and forces that hold these things together. We are just groping through the darkness about how to take molecules and build structures."

What if you could design drugs that, instead of simply controlling hypertension, actually reversed the age-associated damage that causes it in the first place? For some time Pierre Moreau, the dean of the faculty of pharmacology at the University of Montreal, worked in the realm of arteriosclerosis, the vast complex of processes that leads to artery stiffening and, eventually, heart disease and death. He had his eye on one key and understudied form of damage, that known as medial elastocalcinosis, or MEC. MEC is happening to you as you read this. MEC is what happens in your aorta when the little proteins known as elastin, which make it possible for your artery to expand and contract as your heart pumps, come under attack. The invader is calcium, a mineral used by the body for many useful ends, but which, in this case, creates vast damage. Inside the middle of the artery lining, calcium binds to elastin and creates a stiff web of fibers, which turn the once-flexible aorta into a resistant tube. What follows is an increase in blood pressure, and, perhaps just as important, an increase in overall pulse pressure. Both are bad for the heart. No wonder that MEC is now considered a major cause of heart disease.

Moreau, puzzled by why calcium was so attracted to elastin, did what all good medical scientists do today. He created a rodent model of the disease, then subjected the animal to various drug treatments to see if he could affect the process.

What he found was surprising. "At first we were just thinking of using this drug as a preventive against further damage," he recalls. "And that is what we got. But then we were amazed. If you continued the rat on the drug, we found you could remove the calcification from the vessel wall." Like any sober-minded scientist, he has since focused on the mechanisms behind the phenomenon; a human application is some ways off. But the point is right on the de Greyian money. You can remove age-associated damage in a mammal and renew the artery. Monkey with the process—not just to slow it down, but to eliminate age damage altogether.

If you want to eliminate damage to the body, and keep the proverbial human Model T in mint condition, one of the best targets would be excess fat, particularly fat that accumulates around the stomach. This is because we now know that fat cells, or adipocytes, are not simply passive, inert blobs of lipids that look unsightly, but, rather, active, microendocrine organs that, in and of themselves, can do a lot of damage. They spray out all kinds of inflammatory molecules that have been tightly linked to everything from heart disease to brain aging. They predispose us to diabetes. But controlling their accumulation over a lifetime has proven vexing; the human body, after all, was engineered by evolution to deal with food scarcity and in-fection. Until about five hundred years ago, it made perfect sense to have a body that was skewed to acquire and hold on to calories, and to use those cells to fight infection from unsani-tary living conditions. That's what a fat cell was selected to do. Today, in a hygienic, food-abundant world, we do not need it to perform those functions, but it does. Moreover, our entire body has evolved to make it easy to hold on to fat cells. As we get older, it even favors retention of fat cells over retention of muscle cells, something most aging men and women try to fight, with little success, by diet and exercise.

So, what if you could reverse-engineer the human body to fend off fat cell accumulation? Then you would get a body less aged. For the past ten years or so, Professor Kim Janda, a balding, amiable chemist with the jaunty countenance of an athlete, has been working on just such a scenario. Janda, who makes his home at the Scripps Research Institute in La Jolla, got the idea when he was working on another, equally vexing problem—drug addiction. In addiction, almost the same impasse obtained as with obesity. Medical science had come up with various ways to block the effect of various drugs, such as cocaine, but nothing seemed strong and safe enough to use for long periods of time. He began to ask: What if you could immunize the body against a drug molecule, say, cocaine? Using classic vaccine methods, he and his staff made a vaccine from cocaine antibodies, then injected it into lab rats that had been habituated to cocaine. The rats steadily reduced their cocaine use; they had become immune to its gratifying effects.

Janda then started thinking about obesity, and how traditional diets had failed because the body has such a deeply dedicated system of retaining weight. He zeroed in on a hormone produced in the stomach called grehlin. Grehlin's role is to maintain energy balance, and its levels soar when someone diets, sending all kinds of signals through the brain: Decrease energy expenditure. Defend existing weight. Add new weight by stimulating appetite and depressing satiety, the feeling of fullness. Eat that Big Mac. Janda had already seen what would happen if you "knocked out" the gene for grehlin in lab mice;

they stayed leaner, were more active, and tended to burn fat rather than store it. So, what if you made a vaccine that, essentially, immunized the brain from grehlin? Using the same system he used for the cocaine vaccine, he made a grehlin vaccine that targeted the same immune spots in a rat that humans have. The results: less weight gain, more lean mass, less fat mass. As might be expected, Janda attracted enormous interest from drug companies for human trials, although getting permission from a review board to, essentially, *make someone immune to a natural body chemical,* will likely be difficult and require a lot more money.

The vaccine approach permeates experiments in everything from cancer to Alzheimer's, but to date the results have been, at best, mixed. In the case of cancer, it is possible to create a vaccine that prevents breast cancer, but its utility to the aged is limited, because the vaccine's effectiveness depends upon the availability of immune cells, in which the elderly are chronically underfunded. In the realm of Alzheimer's, researchers have been able to create antibodies to amyloid plaque, inject it into humans, and actually clear the plaque from brains. Unfortunately that clearance doesn't translate into clinical benefits; it does not slow or halt dementia onset. Nevertheless, there is some hope that, by artificially modulating plaque-making processes, science will make the brain better able to maintain neural balance, or homeostasis, fending off degeneration and even rejuvenating neural circuitry.

Plaque—all kinds of plaque, from arterial to neural—is one of the great targets of de Grey's "destroy the damage" approach to antiaging medicine, and so it is hardly surprising that one of his more ambitious, and outlandish, proposals takes aim at it not by standard biomedical tactics—reducing cholesterol formation, say, through statin drugs—but via the strange, germy world of environmental engineering. He wants to find a bacterium that eats the stuff.

As de Grey tells it, he got the idea the way he's gotten others, by reconsidering older theories of aging, in this case, the theory of autophagy. Autophagy—"self-eating"—refers to the cell's internal system of waste disposal. Like all cell functions, it is performed by substructures known as organelles; in the case of autophagy these are called lysosomes. For decades, cell biologists have argued about one critical aspect of the balloon-shaped lysosomes, namely, the fact that, with biological age, they become less and less able to completely clean up biological waste—left-behind plaque being the central concern. The debate turned on this: Was the lysosome's declining capacity caused by aging, or was this decline the cause of cellular aging itself? No one could ever agree, and autophagy theory became one of Big G's perennial debate topics, with only a few researchers pursuing it with the zeal of, say, free radical believers.

De Grey, typically, said: so effing what? Who cares if we don't know its ultimate origins? What if we can reengineer or augment the cell so as to keep the lysosome up and running? If you could do that, he reasoned, you could clean up the arterial plaque that causes heart disease and stroke. The same, he reasoned, with certain proteins in the brain and Alzheimer's; if you could restore the lysosome's cleanup ability early in that disease process you could maintain neuronal balance, or homeostasis. Another kind of plaque-like substance known as A2e is a direct cause of macular degeneration. All three of these de Grey rebranded as "lyso-somal storage disorders," a term normally used for a handful of rare diseases. The goal, he argued, was to augment the lysosome and let it clean up the damage. As is his inclination, he asked a hugely unsettling question along the way: If you wanted to find a bacterial enzyme that degraded human cholesterol, where would you look for it? In a graveyard? In a medical waste dump? Where? He talked the National Institute of Aging into sponsoring a roundtable on the subject, and he invited not just the usual cast of biogerontologists, but also all kinds of people who normally did things like clean up oil spills and toxic waste sites.

The project, now dubbed LysoSENS, landed in the laboratory of Pedro J. Alvarez, the chair of Environmental Engineering at Rice University and perhaps one of the nation's leading authorities on bioremediation—the use of soil microbes to clean up waste sites. I met the professor in his office, where we drank strong coffee and Alvarez, an elegant man with an easy Old World way, showed me a series of slides on what has become a lifelong passion—soil bacteria. The field of bioremediation, Alvarez explained, grew out of one central insight by the British bacteriologist E. F. Gale in 1952. Gale, surveying the vast increases in industrial wastes generated during the postwar period, asked a simple question: Why aren't these substances accruing *more* rapidly in the environment? His answer, after endless study of waste sites, was equally simple and even elegant: It was because soil microbes responded to any new energy-rich substance by evolving an ability to use them as a food source; in effect, any new substance created selective pressure on surrounding organisms to use them for food.

This Gale's followers named the "principle of microbial infallibility." It is a theory that has stood the test of time, Alvarez says. "I mean, right now, in my lab, I have soil bacteria that cannot live without the presence of TCE, an incredible environmental toxin that did not exist 140 years ago. Think about that. Look how quickly evolution worked. Remarkable. Really remarkable."

He picked up the thread and went on. "And it occurred to Aubrey, and then to some of us in bioremediation, that maybe it was not such an outlandish idea, that perhaps you could do a form of medical bioremediation." He went on to talk about how the process worked, the endless bench science required to first isolate samples that degraded, say, 7-keto, then identifying the specific gene in that bacteria that coded for the degrading enzyme, and then the long task of figuring out how to get that gene into a mammal, how one might ensure that the gene carried no bad side effects, and how it might then be readied for clinical trials in humans. "Who knows when?" Alvarez said with a twinkle in his eye and a counterbalancing shrug of his slender shoulders. "But this whole endeavor suggests some remarkable things about humans."

It was the second time in the conversation that Alvarez veered into the realm of the "remarkable," so I asked him what he meant. He grew visibly animated. "I mean that maybe we are more plastic as a species than we thought. Remember, until about one billion years ago, horizontal gene transfer was the norm—not through heredity as we have come to think of it, or even through mutations, but from one gene from one organism transferring itself to another for survival reasons. It was how, for example, we got our adaptive immune system." He noted the identification of specific proteins that bacteria use to cross-transfer DNA, the emergence of drug-resistant bacteria as an example, and then pointed to the recent work of several Rice colleagues, who have proven that the majority of DNA in the genomes of some plant and animal species—humans, mice, wheat, and corn—originated from horizontal gene transfer. "I mean, we can actually trace kingdom-to-kingdom gene exchanges." So where does this take us as humans? I asked. "Well, I don't know," Alvarez said, the shrug now triumphant over the twinkle. "I can't, eh, *speculate* on that."

De Grey, of course, can. The last time I saw him we were in Westwood Village, just outside of UCLA, and he was in a good mood. One of his associates had just published a mind-bending paper on autophagy, and how, by improving a certain kind of lysosome process, one could not only prevent liver cell aging, but actually rejuvenate damaged liver cells. In rodents, of course. But there was the proof of principle, modern science's chief fund-raising tool, and if you could just get enough of this kind of work into clinical trials for humans, he said, we might then find it possible to engineer our own "escape velocity."

What was that? I asked. It sounded like some weird transhumanist thing to me. De Grey ordered the duck ragu ravioli and a beer and brushed his enormous beard to the side.

Longevity escape velocity, or LEV, he explained, is what will happen when improvements in "damage elimination" begin to come in regular intervals, each one further extending maximum life span, and each one giving science more time to improve previous treatments. The result, which he has carefully plotted against accepted mortality tables used by mainstream

Big G, would produce life spans of 200, 300, and up to 1,000 years. In the foreseeable future, he sees gains of no less than thirty years. How would that happen? I asked. There would be a panel of state-of-the-art interventions, he said, "maybe every ten years. In hospital, almost certainly, because it's so unprecedented to be doing lots of things at the same time— there would be a big requirement for detailed monitoring of the effects. As time goes on and the treatments become more mature, they will become less laborious; I can imagine them being self-administered in the end."

That reminded me of Pedro Alvarez's comment about horizontal evolution. Was de Grey essentially saying that we will engineer our own antisenescent evolution? "Human evolution will soon be greatly accelerated, despite the reduction in death rate, because of somatic gene therapy," he wrote me later. "I suppose that counts as horizontal gene transfer in a way, but a lot of the changes being introduced will be changes to our existing genes, such as changing our apoE4 [Alzheimer's-prone genes] to [less prone] apoE3."

What would it require to get to the LEV launching pad? The first thing, of course, is money—a huge, war-on-cancer-type effort focused on aging, driven by a combination of enlightened government and private money. That would propel the science to obtain one of de Grey's benchmarks, the expansion of mouse life span by two years, or what he calls RMR, for robust mouse rejuvenation. The second goal was perhaps the more difficult: changing the deeply held beliefs people have about aging. He'd gone on about this before, when we met in Cambridge. "People really go into a sort of pro-aging trance when you start talking about radically extending life. It's as if they'd rather defend something they think they know—that life span is finite—than deal with aging itself as a disease and as something to be defeated." He sipped his beer. "Isn't that amazing? Can you believe it?" I mentioned that it can be troubling when a theory defines all disbelievers as semipathological, but he didn't rise to it. De Grey has become almost too good as a debater.

Just as we were getting ready to go, I asked de Grey how a recent conference he'd organized at UCLA had gone. He looked nonplussed. "Stage three," he said. Stage three? As it turns out, he was referring to Mohandas Gandhi's famous dictum about social change—"First they ignore you, then they laugh at you, then they fight you, then you win." De Grey took it up, not bitter at all. "I've been at Gandhi stage three for maybe a couple of years. If you're trying to make waves, certainly in science, there's a lot of people who are going to have insufficient vision to bother to understand what you're trying to say."

Gandhi? As in the Great Mahatma? I could accept all the talk about SENS, LEV, RMR, horizontal evolution, and the like. But Gandhi-ji? With a shave and a dhoti, he might pass. But, really, had de Grey checked out of the realm of reason, skipped class on reality 101? I mean, was his cookie somehow missing a few chocolate chips? Is beard length inversely correlated with reality? I began to feel a lot like I had in my last encounter with the CR people, or with Thierry Hertoghe. It made me wonder: Does antiaging science inevitably lead one to do a tap dance down the Meshugga Turnpike?

I'd had a chance to run de Grey's vision past a group of highly educated, stylish Londoners gathered at a club to celebrate the birthday of an up-and-coming British comic. They didn't think he was crazy at all, but rather, dangerous.

"It's greed. Just another example of how baby boomers just can't accept limits," one young woman said.

"That'll be good for the environment, won't it?" said a marketing executive. "Just imagine another 50 million people a year on the planet and never fucking leaving! Fucking brilliant!"

"One thousand years! He's got to be an American, right?"

No, de Grey's not an American. But American scientists may be carving out the best middle path to the science of life span extension.

The mole, from Edward Topsell's The History of Four-Footed Beasts and Serpents, *1607.*

BOOK FOUR

The New Longevity Bestiary

Ask now the beasts, and they shall teach thee;
And the fowls on the air, and they shall teach thee;
Or speak to the earth, and it shall teach thee;
And the fishes of the sea shall declare unto thee.

—JOB 12:7–8

The closer we get to truly understanding and intervening in the aging process, the more confusing it all seems to get. Does CR work or not in humans? Does oxidative stress, IGF-1, and hormone modulation really matter? If you look at the science, you constantly walk away with two phrases banging away in your brain: *In mice!* Or *in yeast!* No wonder. At the level of the NIH and NIA, which dole out all the money, there exists today not only a mouse mafia, but a worm mafia, a yeast mafia, and a fly mafia. All of them have vested interests in perpetuating their models, but, as the growth hormone debate and the peculiarities of CR have demonstrated, those models are too simple when it comes to translating the data to humans. Far from the public perception of researchers as white-coated test tube mixers, the real action lies not just with finding the right pill, but with finding the right animal on which to test it.

The Barshop Institute for Longevity and Aging Studies in San Antonio may be the new institutional setting for such work. Tucked into the quiet realm of the Texas desert, the Barshop was founded by Sam and Ann Barshop, philanthropists with, as they tell it, "the ultimate goal of providing humankind with longer lives, free of debilitating disease." I had been hearing about the place for some time and had experienced its vast breadth of insight because, whether I was at a CR conference, a mouse conference, a GSA conference, a lab animal conference, or a neurology conference, I kept coming across Barshop people. Reading across the many lines of its research, what you come away with is this: If we really want to get a handle on human aging and possible interventions, we'll likely have to do something we've been avoiding for some time. We'll have to use a broader array of animals as "models," and that means a renewed wrestling with all the psychic fundamentals we've avoided by using the mouse—now so abstract and emotionally disposable—for so long. In short, we may need a new longevity bestiary, replete with animals not too unlike the one shown in the illustration on page 156.

I'd met Steven Austad, the mustachioed architect of Barshop's comparative biology approach, in the pages of his books about animals and aging, and then, later, at CR and GSA meetings. He is a congenial, cosmopolitan fellow, comfortable at barbecue and boite alike. What struck me, beyond his depth of knowledge, was his continued sense of wonder about the animal kingdom, and, by extension, life itself. He sees aging as the body's "journey through life," and, as he wrote in the preface to his 1997 work, *Why We Age,* "I would hope ... readers might learn to think about aging without terror or tears, to think about it as an intriguing puzzle rather than a gloomy inevitability." He's no advocate of antiaging medicine, be it A4M hormonalism or de Greyian engineering—he's inveighed against both—but he's immersed enough in the dialogue to challenge Big Gerontology's own dogmas, particularly about growth hormone and IGF-1.

For several years now he's been arguing that the growth hormone issue is not as simple as many of his peers think, and that we might be shutting off important reservoirs of innovation by behaving as if it is. He likes to tell a story about hearing his friend and occasional coauthor Richard Miller, one of the brilliant capos of the mouse mafia, talk about low IGF-1 in mice as a determining factor in longevity. Some data had been presented showing that people prone to heart disease because of genes, lack of exercise, or eating junk food did poorly when their IGF-1 was low. "I remember somebody asking Richard if he would rather his daughter have low or high IGF-1 when she was older, and him saying, 'Of course, I'd rather she have low IGF-1,' " Austad said. "I remember sitting there and thinking, well, I don't know about that. Human beings do very poorly with low IGF-1. Very poorly. I don't think we are really reckoning with that, and I don't know why. It may be the case that the mouse is not a good model."

Austad has been intellectually troubled by overreliance on the mouse for some time. To him, the fundamental translational problem with *Mus musculus* as a model for human aging and possible life-extending treatments can be found in one basic measure: weight. More specifically, he is concerned with a ratio: life span for weight. The mouse doesn't track with conventional zoological wisdom. Consider that if one looks across all species of mammals, one observation stands out: Save for a few exceptions, the larger the animal, the longer the life span. A principal missing ingredient in comparative aging research, as Austad saw it, was a way to calculate the longevity quotient, or LQ, of a species—that is, how long a species lived *based on its weight.* That would allow you to compare how long a mouse lives based on its weight with how long a similarly sized, say, bird lives. *That* would be meaningful. Using a vast bank of mammalian data, Austad figured out a way to do so. It works like this: Using 1 as a baseline, a human being has an LQ of 3–5. "They live somewhere between three and five times as long as an average animal of the same size," he explains. They are, thus, slow agers. A mouse, with an LQ of .7, would, theoretically, be a terrible model for human aging. That doesn't mean the mouse is useless, it just means it is limited in terms of its translatability to humans and their vastly different LQ. Remember, he says, the dwarf mouse has yielded up the single most important recent finding in longevity science. "The thing that most amazes me," he says, "is that knocking out a single gene makes the mouse live longer. That was shocking. But it was also complex, because those mice are also all messed up. They are cold. They don't breed . . ." He pauses. "I would never have predicted that."

As with many zoologically inclined thinkers, Austad has a tactile sense of the animal, the mammal unabstracted. Before becoming a professional scientist, he spent years working as a lion tamer (learned from a veteran trainer), as a handler for the actress Tippi Hedren's movie-animal training operation ("I had to leave Hollywood because all anyone talked about was TV. There were no books!"), and even as an animal transporter. (He once conveyed a lion from Oregon to L.A. via the backseat of a beat-up old Mercedes.) Later, as a postdoctoral fellow, he spent years living on a remote island off the Georgia coast, collecting and analyzing possums, trying to figure out why they aged so much more slowly than their mainland counterparts. (The short answer is that, away from predators, their bodies could devote more energy to cellular maintenance and repair functions.)

Visiting with Austad in his Barshop office, which looks out over the forlorn but somehow beautiful Texas desert outside of San Antonio, one gets a sense of where he's headed with all that knowledge and experience. You can see it right outside his window, where tiny desert birds flit by. "The problem with birds is that we can't find any that are short-lived. Even the ones jumping around here live six to seven years. That's longer than any mouse." The same problem obtains with the bat— the mammal with the highest LQ. "We can't get a short-lived one." And it's the same with using primates. Two long-standing studies of CR and rhesus monkeys continue to limp along in one NIH study. That important work is now offering up some tantalizing clues, he says, but modern longevity science needs models that fit into shorter time frames. Hence the popularity of the mouse.

Austad believes there is a midpoint alternative, namely, animals that have LQs that are closer to that of humans and that will, hence, yield up more translatable research results. His dream is a large, multispecies project that would look at aging in everything from naked mole rats to dogs to marmosets, a pint-size new world primate, and, of course, bats. In short, a longevity bestiary. "The nightmare is that there ends up being a hundred ways to be long-lived," he says. "But what we're hoping for is that there's just a handful that would hopefully lead to a limited number of targets for intervention."

Click-clacking away downstairs, in an expanse of clear tubes the size of a small pickup truck, scampers a colony of naked mole rats, exhibit one in the new Barshopian longevity bestiary. The rodents—hairless, ugly, long-toothed, beelike in social structure, and naturally low in several hormones—live about five times longer than they should based on their LQ. To Barshop's new recruit, Rochelle Buffenstein, they represent the future of translational longevity science. Like humans, they age slowly. But unlike humans, they almost never get cancer, their arteries do not get stiff, and they remain reproductively fecund—they *boink*—until the end. Walking me around the colony, and letting me pet the silky outer coat of one animal ("Look at him. He's twenty-eight, and he's just sired"), Buffenstein, who's been studying *Heterocephalis glaber* for more than twenty-five years, paused and traced with her finger a familiar graphic: the rectangularized survival curve so popular these days with antiaging doctors. (And so unpopular with people like my mother when you explain to them that it is the "new Holy Grail.") "See, this is how *we* age," Buffenstein said, tracing an invisible downward slope, "and this is how *they* age," drawing a rectangle. They're perfectly healthy—and then they drop off. "Sometimes I do have to remind myself that they *do* die. And that always brings me back to the fundamental question: Why do animals die?"

For decades now, one of the fundamental answers to that question was: *oxidative damage.* But if you examine the tissue of a mole rat, what you find, she says, is not just a little oxidative damage, but a lot of oxidative damage. In fact, naked mole rats accrue huge amounts of oxidative damage from early on. They simply don't seem to suffer from it. If you take some hydrogen peroxide—a free radical—and put it in a petri dish with some mole-rat cells, the cells not only continue to thrive, but they repair the acute damage more rapidly than shorter-lived animals. To Buffenstein, this suggests that the mole rat's longevity grows out of its ability to defend against acute bouts of oxidative stress. "It's the kind of oxidation that happens because of an unusual occurrence rather than the kind that results from normal cellular breathing." Maybe that means that aging is a "pulsatile" process, not chronic. Maybe the way real aging works is not through long-term processes but by fast, brutal, short-term insults to tissue. The organism that can withstand that kind of stress wins. It's an observation that resounds, not just among gerontology scholars (Finch has even articulated a theory of aging as, among other things, an "event-driven process"), but in real life. One thinks of the way we observe aging in people we have not seen for a while, or in people who have undergone trauma and loss.

In all of this there rests a kind of time bomb for those attached to the mouse-CR model, Buffenstein said. I later asked her what she meant. "As you know, rats and mice have very low LQs, 0.7, meaning they age more rapidly than expected by body size and die quite young, usually from 'age associated diseases' such as cancer," she wrote back. "CR in these animals appears to prolong their lifespan to about that *expected* by mass for those specific species and certainly does *not* extend lifespan to double that expected by body size. As such these may not be the ideal species to examine mechanisms involved in healthy slow aging. But don't tell that to the millions of people studying aging."

Like Austad, Buffenstein comes to the endeavor with a long history of close observational work on mammals. Some of her early years as a graduate student in her native Rhodesia were spent digging eight-foot-deep holes in the hot desert to find a colony. Down there, the eusocialism of the mole rats—they live like bees with a queen and workers—caught her eye. Writing in the *Journal of Comparative Physical Biology*, she noted: "Extended longevity is also correlated with group or social living. Bats that live in large roosts, social primates, colony-dwelling mole rats and honey bees, wasps, and ants all show extended longevity." Why was that? "Enhanced fitness and prolonged longevity in these species may reflect intergenerational transfer of information, communal care of the young, and shared foraging endeavors." They are probably better at cleaning up after each other, and, as a group, they tend to have significantly lower metabolic rates.

Why is that? She grew a little technical for me, but it was utterly worth following the logical flow. "The colony acts as a 'larger individual' with a lower surface-area-to-volume ratio for gas, water and heat exchange with the environment. This lower metabolic rate means that to stay in energy balance the animals in the colony eat less than those housed alone and in effect have a chronic, mild form of caloric restriction. Accompanying the lower metabolic rate and lower need for heat production, mole rat thyroid hormone concentrations are lower

than other mammals." It was a new definition of the benefits
of group living, to say the least. But it also served up a tantalizing clue about hormones and aging. Less, but not none, may
be more, a finding that will likely flummox both antiaging and
anti-antiaging folk alike.

The bat is another one of Austad's many zoological obses-
sions. The second-largest single group of mammals, many bats
are slow aging in the extreme. Yet for decades, bat longevity
was a somewhat closed discussion in gerontology circles, with
the consensus holding that *Chiroptera's* long LQ (9.8) was due
to the fact that it hibernates, and so, as the "rate of living" theory would have it, simply burns fewer bat calories and makes
fewer bat free radicals. Then data started coming in from field
studies of tropical bats that don't hibernate. That blew everything up, Austad says, perfectly delighted. "They lived just as
long!" What's more, the bats in the field studies showed the
same kind of resistance to age-related pathology as CR mice.
"The Brandt's bats that lived thirty-eight years had to fly more
acrobatically than their insect prey for nearly four decades,"
he marvels. "Over that time, it had to catch enough prey to
support its high-energy metabolism during flight as well as
successfully escape its predators and pathogens. And it had
to preserve its voice and acute hearing for decades in order to
localize its prey in the dark using its echolocation ability." He
theorizes that the bat, like the mole rat, may have developed a
kind of evasion metabolism, one that shunts energy into repair
and maintenance rather than defense, growth, and reproduction. If that is the case, can we locate a gene or genes for such?
Again, bats offer another longevist's dream: Because they are
not *all* long-lived, it would be possible to compare the genomes

of long- and short-lived species within the same order—and that, Austad says, might pay off someday in more relevant targets for human longevity therapeutics.

One day, between various staff and faculty conferences at Barshop, Austad paused for a moment over his desk and took in the cover of an issue of *Science,* which showed a large dog and a small dog. He slapped it down on his chair and chuckled in the manner of a man with many passionate—and vexingly unanswered—questions. "I want to know why *that* happens," he said. "Why is it that the small dog lives almost twice as long as the big one, and with a lot fewer pathologies, to boot?" It is one of the zillion-dollar mysteries in human aging as well, for if it is generally true that, between species, the bigger the animal the longer the life span, it is also arguably true that, when you compare members of the *same* species to one another, the smaller are often the more long-lived. There is even a famous study of professional baseball players indicating that the short guys lived the longest. Yet the most robust data of this phenomenon are not found in mice or rats or humans or horses, but in our best friend, *Canis.* Austad, whose wife is a well-known veterinarian with whom he once coauthored a book, draws from a huge bank of breeding and veterinary data to note, for example, that the age for large dogs to begin geriatric care ranges from six to nine years, whereas smaller breeds need not commence the proverbial Lipitor until they are anywhere from nine to thirteen. Why is that, and wouldn't using the well-kept canine breeding data and the recently completed dog genome get us a great deal closer to human-sized aging mechanisms?

Yet just as in a discussion about weight with one's wife or

husband, you can get in all kinds of trouble when you introduce the element of size in any analysis of aging. Austad is
constantly coming up against a kind of anti-sizcism, to steal a
term from the professional fatties. "Elephants contain about
forty times the numbers of cells we do, and whales as many as
six hundred times as many cells. Yet elephants and whales live,
to a reasonable approximation, just as long as we do. Therefore, their cells must be forty to six hundred times *more* resistant to turning cancerous than our own. Could we perhaps
learn something about cancer resistance from studying these
cells?" Whenever he mentions this at conferences, he says, he
gets a lot of guff.

The mouse, of course, still commands huge utility in the
longevity bestiary, especially when you switch out of the
CR realm and into the arena of nutritional supplements. One
tack is to use the mouse to debunk claims of antiaging huck-
sters. In recent years, melatonin, coenzyme Q10, L-arginine
supplements—all have come under the Barshopian mouse
lens and been found wanting. For years it was institutional
sport. (As one researcher told me, "Someone would come in
with something, and we would say, oh, give it to [Jim] Nelson
[another Barshop scientist]. He'll shoot it between the eyes.")
What has been surprising in recent years are the number of
compounds, scrutinized in tandem with the Miller Lab in
Michigan and the Harrison group at Jackson Laboratories in
Maine, that seem to extend maximum mouse life span. "This,
to us, has been a huge shock," Randy Strong, who runs the
testing program, told me. "I mean, we are all pretty die-hard
skeptics here. When we designed the program, it was basically
with the mind-set that we would be in the business of finding

out that none of this stuff worked." Among the elements that extend mouse life span significantly are a compound of creosote (unfortunately a liver toxin in humans) and something called rapomycin, part of an established aging pathway that only recently received renewed attention.

Mole rats, bats, dogs, and mice—it seemed to me that, as go-ahead as all those models were, we were still dancing around the tender issue: Where were the primates? Only primates get you really close to some issues in human aging and, by extension, possible interventions. As Finch once chortled in a discussion of better animal models, "What you really need is a dwarf monkey that only lives five years!" Suzette Tardif, another recent Barshop recruit, may have something close. You can see it on her office wall. There hangs a photo of two young mammals—Tardif, when she was a high school science geek thirty years ago, and, opposite her sunny face, perched on her hand, a marmoset, the dwarf monkey that has been her lifelong object of study and contemplation. The common marmoset, or *Callithrix jacchus*, lives about seven to fifteen years, with an LQ of 1.8. That's not quite the neat package Finch dreams about, but, as Tardif says, "If you want to be flippant about it, the marmoset lives about as long as a completed NIH grant." Small note: it is not an endangered species.

Tardif acquired her passion for this New World primate as a summer intern at the Oak Ridge National Laboratory in Tennessee. Back then, there was interest in figuring out why marmosets seemed to be resistant to radiation, but "that work went nowhere," she says. A veteran researcher then got her interested in marmoset breeding, trying to figure out why the animals were so hard to establish as a colony. That got her hooked. What followed were stints at the Smithsonian and the Southwest National Primate Center, where she looked at some of the hottest biomed topics of the day: in utero programming, obesity, and diabetes, all in the marmoset.

What guided her toward aging was energetics—how the animal partitioned its use of fuel, the same obsession surrounding almost every inquiry into mouse CR longevity. The big unanswered question to her and Austad was: how does that translate to humans? How can we find out what altered energetics means when it comes to examining why, say, an aging person loses muscle mass and replaces it with fat? Why do we become frail? Is it a random process or one with identifiable interventions? You can't really do that with a mouse. Here the marmoset, closer to us in terms of LQ and evolutionary history, might yield up some long-awaited, and perhaps even therapeutic, answers. Other scholars have shown, for example, that if you look at how a marmoset's joints age, in the jaw, you find nearly identical bone cell processes. If you look at how the marmoset's retinal cells degenerate, you find more similarities to that of humans. Tardif already has linked the decline of marmoset thigh circumference to age, something familiar to most aging humans. And that will inevitably bring her back to deal with one of the big scraps in Big G: IGF-1 and growth hormone. With it, you can prevent mass muscle loss—sarcopenia—a big cause of disability in older humans. "Yes, low IGF-1 in transgenic mice makes them live longer. But for humans it has a cost. Why we are not dealing with that dichotomy more directly I do not know. It seems that the people studying the molecules do not want to deal with that." They don't want to look at the body. Like her fellows, she has no truck with the A4M crowd either. "The research environment of aging is one in which you can become very rich in antiaging," she said. "There is pressure to overstate, to put it mildly."

Driving me back to my hotel one evening, Austad pointed out the local primate center, which you can see from the freeway. It set us both to talking about the great unsaid in all discussions of animal research: How humane is it? Can it be made more humane? Will society support such work—involving, as it invariably does, unnatural confinement, testing (however humane), dissection (sometimes long before the animal would normally die), and, of course, euthanasia. If you spend any time with animal researchers, you generally find that the answer to this falls into one of two categories: silent pouty rage, later accompanied by some form of the exhortation "But they would never live so well in the wild!"; or salutary but empty gestures of empathy. Austad was different. His was the view of the pragmatic zoologist, someone who still wonders at animals but, perhaps having handled them so closely over the years, has few delusions about their essential humanness. Moving upward from mice will actually encourage better animal husbandry. Dogs and marmosets, after all, are much better regulated than mice. And look at Tardif and Buffenstein—they had taken their colonies with them over decades of work. You're not going to get better attentiveness to healthy animals than theirs.

Is America ready for a new longevity bestiary? Peruse any popular discussion about laboratory animals, filled as it inevitably is with invective, moralization, anthropomorphism, and downright intellectual and physical bullying, and you have to wonder.

Then again, you also have to wonder if we're ready for one million centenarians. . . .

Lucas Cranach's Fountain of Youth, *1546.*

BOOK FIVE

One Million Centenarians

[Regarding the nutrition of older people], if I had to choose between the whole science of modern biochemistry and the wisdom of the past centuries based on human experience I believe I could guide people to better advantage by the use of the wisdom. —CLIVE MCCAY

Assume, for the moment, that de Grey, Hertoghe, and the mouse mafia and its minions in CR aren't crazy, but, rather, that a combination of their methods, perhaps made more applicable by something like a new longevity bestiary, adds one year to life expectancy every year for the next twenty years. That is not an extreme assumption. If you look at the work of people who get paid to model the future, it is considered a moderate assumption, something like de Grey's 50 percent chance. Shripad Tuljapurkar, Stanford University's dean of population studies, undertook such a modeling study a few years ago; it confirmed much of what he has been observing ever since the late 1990s, when he began studying aging societies around the world. It's this: Just as investments in childhood health in the twentieth century brought down early-life mortality rates and so boosted average life expectancy, investments in delaying aging will result in similar increases in life expectancy. This, he says, brings with it all of the problems you might imagine: an increase in government spending, a strain on existing medical resources, inequalities in care, and an irrational boom in demand for pirated versions of the Grateful Dead's spring 1972 concert in Frankfurt whereJerryplayedthatawesomefifteenminutesolo on "Truckin'." As Tuljapurkar told the annual gathering of the American Association for the Advancement of Science in St. Louis, Missouri, in 2006: "People are going to do things they didn't get 'round to in their working lives. Current institutions are really not equipped at the moment to deal with such long lives." In the United States, there are currently 90,422 centenarians, compared with 50,424 in 2000. By 2050 there could be as many as 1.1 million.

A million centenarians.

But what if such long lives came without, or at least with a lot fewer, disabilities? If you erase one important but, by most accounts, overplayed notion—that obesity will slow down our two-century gain in life expectancy—you come away with a completely different view than that of Tuljapurkar: We will delay decrepitude. We will experience not just longer lives, but those added years will be healthy years. On this score, what amazes is the relatively positive nature of some of the biggest antiaging skeptics. Take them and put them in a room and ask them to estimate the chances over the next twenty years of major antiaging therapies, and they tend to include things like this, taken from a RAND study in 2005: likelihood of a selective estrogen receptor modulator that will decrease breast cancer by 30 percent, 90 percent; of a telomerase inhibitor that will cure 50 percent of all cancers, 100 percent; of an outright cure for half of all metastatic disease, 70–100 percent; of a treatment for Parkinson's via neurotransplantation, 25 percent; of primary prevention of Alzheimer's using antiamyloid therapies, 40 percent. There is optimism ("0–50 percent likelihood") on extending maximal healthy life span too, RAND's skeptics said, which would mean "10–20 years of extra life of an equivalency between twenty and fifty years of age."

Optimism, yes, but what of wisdom?

Wisdom, being in short supply in almost all arenas of modern life, seems particularly short in the realm of aging, antiaging medicine, and longevity science, but there is one man who seems to have some of it under his hat. That man is Robert Butler, the dean of modern geriatrics, the founding director of the National Institute of Aging, and the man who has done the most over the decades to fight discrimination against the aged. In the late 1950s, as a young physician investigating nursing home abuse, he made a slew of enemies by publicly concluding that "I saw neither nursing nor homes." It was a Twain-like cathartic that spawned a movement. Butler later detailed the abuse, and problems of aging in the United States, in a book entitled *Why Survive?*, which won the Pulitzer Prize in 1976. In recent years, he cleaved to the role of senior statesman, with, to my mind, an understandable reactionary tendency. He was one of the name signatories to the broadsides against de Grey and SENS, and against A4M and antiaging medicine in general. In short, his tenure had witnessed the transformation of aging into a marketing opportunity, and, as is perhaps appropriate in a market-mad culture, his role was increasingly one of buyer beware. The tendency had extended to brochures in his office warning against antiaging medicine, where the visitor was appropriately warned.

I was surprised, then, when I began to see Butler's name pop up in forums about longevity and extended life spans. In 2008, he was on TV talking the subject up with Barbara Walters. He was a big supporter of the mouse, yeast, fly, and worm mafias. He had a huge and often lively new book, *The Longevity Revolution*, and in it appeared a number of startling statements for a fellow who had not long ago dubbed SENS a farrago and A4M outright quackery:

> • *The present level of development of aging and longevity research justifies an Apollo-type effort to control aging. (p. 187)*
> • *. . . in the twenty-first century we may be poised at the frontier of biological time—the prospect of germline engineering, the means by which the human species would direct its own evolution and extend its life span. (p. 401)*
> • *. . . it may soon be possible to delay both aging and age-related disease in humans. (p.162)*
> • *One must not doubt the possibility of the unexpected in science and the uneven evolution of knowledge. (p. 401)*
> • *Advances in genomics and regenerative medicine will give us advances in longevity. (p. 298)*

An Apollo-type effort to control aging. The frontier of biological time. Directing our own evolution. These were not the carefully gauged words of a statesman. These came from the lexicon of vision and transformation. What had brought that about? Perhaps it was Butler's own aging. He had just turned eighty, and he had been ill; before a GSA speech the previous year he was bemoaning to friends that he was due for kidney surgery, and he looked haggard. Only a couple of years before he'd lost his wife. The loss had been devastating and the loneliness, he said, had begun to wear. The pragmatist in me urged me to visit with him sooner than later.

At his Upper East Side office at the International Longevity Center, which he founded, Butler, looking a lot better than the last time I had seen him, was aloft on the issues of the day. A new study on geriatric training for medical students (there basically isn't any) brought NBC news to his door. Another report, on nursing homes, was pushing his favorite topic to the fore again. "It is amazing," he said. "*Still*, only one out of ten homes meets federal standards." Was it all some sort of insidious new form of ageism? After all, that very day there appeared a front-page story in the *New York Times* detailing how insurance companies were limiting the choice of key drugs for chronic diseases. Buried in the story was the fact that the same thing had been happening to the elderly since the new prescription drug act had taken effect three years before. Butler lifted his hands in mock resignation. What more could he say?

I asked him why he had turned the corner on antiaging, or longevity, medicine, and its potential. Unprompted, he lectured me for a few moments on how he had "not attacked anyone personally," and that he had a duty to call attention to the more extreme claims of commercial antiaging. But he conceded that the realm of what he considered possible had indeed expanded. "It's the science," Butler said. "It's been getting better and better, and with the right kind of strategic investment, we may well come away with some therapies that significantly expand life expectancy and even life span." Thus his plan for a new Apollo-like program for longevity science. "Today less than 1 percent of the entire federal budget is spent on medical research," he said, quoting from his book. "We could dramatically improve health and control costs if we took 3 percent of the nation's overall health bill, which would work out to $54 billion, and make it available to the NIH for medical research. I also propose that 1 percent, or $3 billion, of Medicare expenditures be devoted to the National Institute on Aging." What about growth hormone? I asked. "We are ready for prime time on that," he said, his eyes twinkling a little impishly. What did he mean by that? "I mean that we are ready for large clinical trials. Big ones. And also more work to find a version of HGH that confers the aging benefits without the risks."

That was his science package. Getting his head around the more practical issues of a longevist society has taken more time; in the past, Butler concerned himself with showing, in public policy debates, that the old do not consume disproportionate health care dollars. In the new longevity revolution, he had to retool and try to answer a more uncomfortable question: How do we pay for an extended-life-expectancy society? By all accounts it will be more expensive. Butler's answer is typically hardheaded and frank, something you don't tend to find in many of the rosier how-to "manuals" of modern aging. We will simply have to work longer, he says. The government will have to encourage industry to delay retirement, extend flextime, and invest in serial retraining. Keeping more people in the tax-paying workforce will shore up Social Security as well. There would be new inducements to save money through new forms of life annuities accounts. All of this, he said, "would constitute a kind of neo-welfare state—a new covenant—that promotes individual responsibility in alliance with the voluntary sector, the market, and government. But a secure, tax-based safety net must still back up such a reform." He liked to call it "productive aging."

Given the nature of American political leadership, I suggested that this might happen right after pigs sprouted wings, but Butler was undaunted. The cold water of retirement was about to splash the faces of the boomer generation, he said, and when that happened, things will begin to change, perhaps not for the boomers themselves, but for the following generations. (The boomers, unfortunately, would remain "a generation at risk. They are simply not prepared for it all.") He pointed out that the idea of productive aging—the latest version of successful aging—is hardly a new one. The Russian Nobelist Ilya Metchnikoff, perhaps the progenitor of modern geriatric thinking, once predicted that "the prolongation of life would be associated with the preservation of intelligence and the power to work . . ." Ideally, Metchnikoff went on, "when we have reduced or abolished such causes of precocious senility such as intemperance and disease, it will no longer be necessary to give pensions at the age of sixty or seventy years. . . . We must use all our endeavors to allow men to complete their normal course of life, and to make it possible for old men to play their parts as advisers and judges, endowed with their long experience of life."

Some of Butler's agenda is already under way, naughty Americans not waiting for an official movement or new legislation. The modern newspaper obituary, increasingly peppered with ages like 91, 95, 98, 101, tells the story better than any novelist. There is the schoolteacher who, having retired, took up inventing and started a thriving business; the technology executive, downsized at sixty-five, who took up mentoring of local youth and now runs a charter school; the bank manager, pensioned off at sixty-two, who began to paint; the housewife, widowed at seventy, who raised her daughter's children and started a thriving preschool in the process. One wonders where Americans might take it all.

We will also deal with something else. With more time, both on our hands and in total, we will experience more — more change, more distance from the young, more distance from the present, and, given human nature, more attachment to the past. In a sense, we will enter our own dimension. It is not something you tend to hear much about from the big thinkers, but if you get lucky, as I did not once but twice, you get a sense of it.

The first time came while taking a graduate course in the biology of aging at the University of Southern California. The professor was none other than Caleb Finch, the evolutionary biologist who, among many other things, coined the term "negligible senescence." I had gotten to know Finch, who travels by the nickname Tuck, during my wanderings in various Big and Little G circles, from CR to SENS to the annual GSA convention, and he had come to be a consistent source of insight, camaraderie, and intellectual challenge. (As was once said of the physicist and Nobel laureate Murray Gell-Mann, Finch has two brains but only uses one to keep up with most of us.) The class met every Monday and consisted of grad students and postdocs mainly in their late twenties to midthirties. For weeks we had buckled down and inhaled hundreds of pages of journal articles on the minutiae of aging, from dysfunctional lipoproteins to glucose disposition indices to delayed fruit fly reproductive schedules, and listened to the cream of Big G's current crop of evolutionary cell biologists, all of whom were there because of Finch. It was an overwhelming, sometime narcotizing, feast of data, with at least one grad student loudly banging his head on the table when he nodded off, much to our collective delight.

Then, late in the term, the autumn afternoon turning all gold through the windows, Finch took the podium. He is, to put it mildly, an imposing figure, tall and bearded with a thinker's forehead and a querying countenance. Quickly he moved through his classic discussion of inflammation, its evolutionary origins, and its purchase on the modern human aging process. It is, as he himself will admit, a demanding presentation—a relentless flow of data and logic, data and logic. One can feel the limits of having only one brain. As he came to the end of his lecture, Finch flipped off the PowerPoint and gazed out at the class, as if he were looking for a particular kind of flower in large field of wild grass. "Now, hmm, there is one other thing that we haven't talked much about that is a huge factor in aging. Can anyone tell me what that might be?" There was a silence. Then, first from one student, then from another and another:

"Macrophage infiltration and foam cell formation?"

No.

"Glucose-protein matrix accumulation?"

No.

"Autophagy!"

No.

"Sirt-1 alleviation of endoplasmic reticulum stress?"

Finch drew a long breath and looked down at his hands. "No . . . that . . . wasn't exactly what I was looking for. No one?" He paused. "It's . . . loss."

Loss?

"Loss—that's right. As a human being gets older, that person is inevitably going to experience loss—the death of someone close, the end of a relationship, a career. And as that happens, it . . ." He paused, dead serious, the room utterly rapt. "It takes it out of you. It changes you. Something is different."

Perhaps it was the entire effect, the dry presentation combined with the humane, almost emotive counterpoint, but for the first time in the term, something seemed to break loose. The talk afterward was animated, as if everyone finally had made a connection between molecules and man. Finch had put his finger on something that had always bothered me about life extension. In short, how will we best spend this time, a time of unprecedented health but also a time almost completely unexplored, psychically and psychologically, by previous generations? It would be a new dimension. Was there anyone who had explored it? And, I'm sorry, how would we survive all the bad art that will flow from it? Again, I got lucky.

On September 2, 2006, a small article appeared in the back pages of the *Los Angeles Times*. It was an obituary about a man named George Johnson. Johnson, the paper said, had been the oldest World War I veteran living in California. He had just died of natural causes—at age 112. What was immediately interesting was Johnson's lifestyle, of which everyone tried to make something: He pretty much ate what he wanted; smoked, he had said, "but only until I was 70"; and, perhaps most remarkable, worked for forty years in the asbestos-laden Kaiser Shipyards outside of Oakland, California, as a furnace man. As is the case with almost everyone who displays what experts now term exceptional longevity—over one hundred—Johnson's many years likely had more to do with genes than his choice of breakfast cereals. Something—a mutated growth hormone receptor, an altered insulin pathway, a highly attuned network of anti-inflammatory genes, who knew? I asked his great-nephew, Brian Johnson, if I might see George's home and attend the funeral, and I drove up to the old bayside enclave of Richmond, where Johnson had lived for nearly seventy years.

Brian Johnson met me at the door. He was a congenial man, himself a retired insurance executive, and we quickly fell to talking about the seemingly endless flow of longevity-themed news stories concerning his great-uncle. There was the requisite "ironic": Johnson "ate nothing but waffles and sausages" (Associated Press). There was the "Lincolnesque": Johnson had "built his own house from wood scraps" (*LA Times*). There was even a temperance spin: He "had only tasted liquor once in his life" (Public Radio). In all of these there was much truth, but there was also "a lot of bullshit," Brian said, "courtesy of my uncle." He nodded toward a small, quarter-empty bottle of inexpensive brandy sitting on a shelf nearby. We laughed.

Arriving the day before Johnson's funeral, I'd found the good nephew and his wife diligently and deliberately plowing through the vast ephemera of George Johnson's long life. "Look!" Brian proclaimed at one point, holding up some small notepapers. "I think he had a bookie!" His wife mock-frowned at him. Johnson looked over at me and shrugged. "I'm constantly amazed at all this. I mean, here I am, planning a funeral for a man who was not ill, was not injured, and certainly did not die a premature death. It's hard to believe!"

The surprises had been coming for some time. Brian Johnson did not know he had a great-uncle until the 1970s, when his grandfather Walt—George's brother—died. The family history gap was itself a remnant of early-twentieth-century social convention: the nondisclosure of mixed racial heritage. George and his brother Walt were half African American, half white, and the two brothers had early on agreed that, in order not to "burden" the next generation of Johnsons, they would not tell any offspring of their mixed-blood heritage. It was a huge family secret, requiring no end of dodges and evasions, but it all came out at Walt's funeral thirty years ago. "When I found out," said Brian Johnson, "it broke my heart." He held his hands to his chest and his eyes welled as he spoke. "I mean . . . I'd never heard of an Uncle George."

Brian then began to make trips to his lost uncle's home. It was the 1970s, and while both George and his wife, Ida, by then had been retired for over a decade—he from Kaiser, she from long stints as a cosmetologist and a restaurateur—the couple's life was that of constant, albeit calm, social activity. There was always someone staying overnight, or someone's nephew living with them temporarily. "They were always doing something in the house—and not watching TV," Johnson recalled, throwing a glance at the mid-1960s RCA-Victor with rabbit ears next to him.

George and Ida's house was divided into three identifiable realms of activity: the social, domestic, and the avocational. George built the home himself in the 1930s, when he was working at Kaiser. "Every single board," he liked to say of the effort. "Every single nail . . ." Although the source and type of the structural wood is unclear, much of the interior is plywood, some stained a muddy red, perhaps to match the outside of the structure, which George covered with a mottled rose stucco. From the street the house stands as pure working-class efficiency, three floors high, a perfect bump in the gray, working-class knob of a neighborhood where it sits. When the fog on the bay clears, the view is all gold and sun. "I can't believe that this will be a teardown in this market," Brian Johnson said. "But that's what it is."

We walked up the stairs to the third floor, a large, single room with a grand picture window view, the kind you couldn't install these days because it would be out of code. "This was the party room," Johnson says. "There was always something going on up here." He left me for a while to deal with a Realtor who'd knocked on the door downstairs, and he invited me to scrutinize while he was gone. The walls of the party room stand mostly bare, save for a few faded prints of Venice and Vermeer's *Girl with a Pearl Earring*. The bric-a-brac was minimal; there is a fifty-year wedding anniversary ashtray with gold bells attached.

The main attraction of the Johnsons' social realm, besides the view, was the reading material. This was a household of self-educators. Books and magazines abound. Almost all of the latter date from the 1960s and 1970s, most of them general interest, as they were once known, with cover stories on "Charlie Chaplin and Sophia Loren" (*Look*, 1966), or on "Pearl Buck's Cookbook" (*Ladies Home Journal*, 1973). The book titles conjure up a journey through American twentieth-century taste, from early popular novels (*Husbands and Horses*), to depression-era high lit (*Tender Is the Night*), to instructional manuals (*First Course in Algebra*), hobby lit (*Uranium and Fluorescent Minerals — A Guide*), interpersonal self-help (*Let's Start Over Again*), off-beat Christian (*Origin and Evolution of the Soul According to the German Mystics* and *Social Christianity*), and early postwar New Age (*The Power of Rhythmic Prayer* and *I AM*). Ida, who was the cook in the household, compiled a compendium of early Weight Watchers recipes and menus and had them bound together.

Down on the bottom floor—the avocational realm— George repaired things as the house and he grew old. "He drove this car until he was 102," Brian said as we made our way into a cramped garage and workbench area and took in the sight of George's 1963 Chrysler. "He was constantly fixing something, and I'm convinced that his do-it-yourself bent has a lot to do with him living so long. He saw everything as a potential expense that he could either avoid by fixing something, or by doing without it. He was on a fixed pension for forty-two years and he was not going to take out a second mortgage, as conventional wisdom might have suggested! So he was constantly trimming his own tree, or fixing his own water heater. I mean, here was a guy who only heated the room he was in, to save money. One of his few concessions to technology, at the very end, was to let me order his groceries by Internet."

A 1963 Chrysler and the Internet—it was between those inventions that George Johnson spent most of his retirement. Commenced in 1964, the years were largely spent on the second floor, the domestic realm. (During his last ten, he went blind.) The domestic realm was, by modern standards, technology poor in the extreme. In the kitchen, Ida's old gas-top stove, circa 1940 and barely used in recent years, got supplanted some years ago by a single burner hot plate, where George could warm his hot dogs and sausages. "He liked those because when meals-on-wheels was not around he could make his own snacks, and because he knew every inch of this house he could pull that off even being blind," Brian explained. Up on the shelves were convenience foods like Rice-A-Roni that, near the end, George's careworker, a gentle Tongan woman, could heat up for him.

"This is where he listened to the Giants games," Brian said as we entered George's living room. "He was a huge fan." In the corner of the room, just below another spectacular bay view, sat George's improvised lounge chair, covered with a hand-knotted wool shawl. His headphones, still plugged into an eight-track deck with his favorite music plugged in (the Mormon Tabernacle Choir), hung from the side. On his bedroom door hung a handwritten sign—GEORGE'S ROOM KNOCK LOUD AND CALL GEORGE—and inside, Johnson's low-tech, zero-sum ethos continued: there was no Celebrex, just Ben-Gay; no prescription stool softener, just an enema bag; no incontinence pill, just a box of underwear liners. An ancient Dimplex electric floor heater warmed him on particularly cold days. Plastic flowers sprouted here and there, and in perhaps his one nod to race consciousness, George had hung above his bed a crucifix *and* a nubile art deco Cleopatra.

Change and loss. Change and loss. George Johnson seemed
to float on that sea, with no signs of dementia, and without
what we have come to consider the comforts of modern life.
He lived poor, but not poorly.

And so: Constant activity. Self-sufficiency. Living low. A
slightly enhanced tolerance for coldness and the Mormon
Tabernacle Choir. The first two would come as no shock to
geriatricians as examples of what they like to call "successful
aging." But those last two—living low and cold tolerance—
that sounded to me like the phenotype of the classic CR mouse,
or, for that matter, CR person. A reasonable layperson might
thus ask: Was there an identifiable George Johnson longevity
phenotype—did those outward traits and behavior register in
his body and contribute to his longevity? In fact, was there
anything remarkable at all about his body?

It is rare for forensic science to bother with such questions,
what with stabbings, poisonings, and gunshots being so much
more dramatic and potentially telegenic, but, in the case of
Johnson, the world was lucky. Johnson had consented to an
autopsy, and a few hours after he died he was ferried to the
Palo Alto VA hospital, where a team of pathologists examined
him in detail. The effort was coordinated by Stephen Coles,
an adjunct professor at UCLA Medical School and the
founder, president, chief fund-raiser, and scientific director
for the Supercentenarian Research Project. Coles started the
organization to track the lives and deaths of "everyone on
earth who lives beyond 110 years old," as he told me one day
in his tiny, journal-packed office on the southwest side of Los
Angeles. "We want to have a precise, comprehensive database
about these remarkable people. We can't leave it to memory,

to chance. This is important." Coles had given me the same purpose-statement lecture twice already, at two different conferences, but, professorially fuzzy-headed at times, he frankly didn't recall me at all. As an example of just how concrete his pursuits had grown, he motioned to a small jar sitting on top of his office microwave and said, "That's his pituitary, you know." Inside a baby food jar–sized beaker floated a pea of reddish brown flesh. "Johnson's."

I experienced my usual moment of quease, stifled an impulse to make sure all my vital organs were adequately protected, then asked him what the autopsy had discovered. Coles shifted the conversation to Johnson's aorta. "Amazing," he said, pulling up slides of the examination on his big screen computer. There, displayed in the sterile laboratory setting of modern medical science, was George Johnson's aorta, flayed and opened, pink and tender. "Look at that," Coles went on, clearly excited. "It is perfect. There is no plaque, no inflammation, it's like he was twenty years old. Why is that?" Unfortunately for me, that observation stimulated a thirty-minute dissertation on stochastic damage and genetically driven resistance to molecular infidelity and other gero-cocktail chatter. When I recovered, Coles had gone on to Johnson's poor little pituitary. "Again, remarkable, but this time not because it is perfect, but because it was . . ."—his eyes brightened— "to-ta-lly a-cell-u-lar!"

A totally acellular pituitary?

This rang a reasonable mouse bell: The pituitary is what stimulates growth hormone. Low growth hormone is associated with maximum life span extension; that is—as Andrzej Bartke would scream if he weren't such a gentleman—"*in mice!*" I wanted to know more. What did he mean by acellular, and how common was this in autopsies of average elderly people? Coles didn't know.

I asked Dr. Harry Vinters, a cellular and molecular pathologist at UCLA, to take a look at the slides. At Johnson's aorta, he too marveled. "That kind of presentation—really almost no pathology—is very rare," he said. It is worth studying. What about the pituitary? There Vinters's read was disheartening. "See," he said, clicking the slide up on his big screen computer, "this, for one thing, is a slide of the *wrong part* of the pituitary. This is the posterior lobe—it doesn't secrete growth hormone. And, frankly, this is not so unusual to see this." A few weeks later, we got the Palo Alto Veterans Hospital to send down slides of the growth-hormone-secreting part of Johnson's anterior pituitary. I drove across town to get a one-sentence read from Vinters. "Sorry," he said, "it's totally unremarkable."

Driving home, I thought about Johnson's longevity again. What else was remarkable? One thing was the *nature* of his and Ida's extended social life. The more I spoke with friends, relatives, and neighbors, the more it became apparent that the Johnsons possessed something rare: Even in their nineties, their circle of friends and intimates included lots of young people. "There was always someone staying with them, like, for a semester, or even for a whole school year, going to Berkeley or something," one neighbor recalled at Johnson's funeral. "When I was staying with him, in my twenties, there was always something going on upstairs, some cocktail party, some dinner, something," a great-niece recalled. "It was just part of their world, natural-like."

It isn't so natural-like these days. Age segregation—both self and socially imposed—have become the norm for the aged. Self-driven segregation you can see in places like Santa Fe, New Mexico, now a kind of gero-art colony, a place where the old person's main contact with the young is with the service class. You can see it in places like Palm Springs and Rancho Mirage—God's waiting rooms. There the principal contact is with the visiting grandchildren. The kids like the swimming pool and Gramma a lot but hate the weather and the long drive to get there. It is hardly an American phenomenon, either, as anyone who has visited elders in Italy and Japan can attest. In the former, the main contact with the young is with a Polish caretaker, often a twenty- to thirty-year-old who barely speaks the language. Even in the developing world, societies are getting older before they get richer. Fertility is down worldwide. Gains in population come increasingly from extended life expectancy—more older folks. The norm is socially driven age segregation, principally of economic origins. Both kinds display signs of loneliness and alienation. In America, if you open your eyes and heart at all, you can see and hear it everywhere.

It's true that America, in its straightforward, businesslike way, is also slowly evolving its own longevity social phenotype: Exercise and yoga, constant learning, intentional conviviality—all of these are part of the modern recipe for successful aging. But the more I speak with the elderly—everyone from my own mother to former academics now in retirement—the more I tend to hear one thing: They want to be around young people. But young people don't really want to be around them. How can we make long life less lonely, less attached to the past and more toward the now, the kind of constantly updated dimension that the Johnsons so economically created for themselves? Cornaro had one solution: Have the grandkids take up residence under your own roof. That, of course, was part of the great signorial tradition, and a more acceptable version of the most ancient of all geriatric advice: To stay young, sleep with virgins.

Here, then, is this book's only policy recommendation for our new, longevity-enhanced world: We must find a way to subsidize young families to live in older communities. I can already hear the various vested interests screaming about that one, but I really don't care. If you want to honestly deal with the psychological issues of a healthy older population, there's one that cuts across all the lines—class, race, gender, and even politics. Policymakers—especially you burgeoning corps of urban planners—figure it out.

There are all kinds of other reforms and programs one might advocate—Butler's plan for an Apollo effort seems about right for the science side, especially if it is buttressed by something as hardheaded as Austad's bestiary. Some kind of regulation of human growth hormone—perhaps requiring compounding pharmacies to collect side-effect data—would be helpful. We need to underwrite the training of a new generation of biogeriatricians, physicians who understand how to proactively treat the signs of aging. And, of course, we need to figure out a way to pay for it all. As Butler said, these measures might not help us directly, but they would make our children's lives longer and healthier and more comfortable.

Yet in the end, my own testosterone steadied, I kept coming back to Alvise Cornaro. If you travel to Padua, where he made his home and society, you can, as I did, get some sense of that aging man and mind. In parts of his house, now restored, you can walk in the courtyard that once hosted the great thinkers of his day. In his vaulted music room and classical loggia, or theater, you can put you head back and gaze up at the intricately decorated ceilings and frescoes, and you can see what Alvise, at sixty, seventy, eighty years of age, saw: grotesques from the classical period, half-nymphs and half-horses, colorful versions of the ultimate changeling, the goddess Diana. One can walk up the street and stand where Vesalius held his first public dissections, in the process exploding everyone's old notions about the human body and why it functioned the way it did. (This at about the time that Cornaro first contemplated *La Vita Sobria*.) About a five-minute walk away stands the Orto Botanica, one of Europe's first botanical gardens, built by Daniele Barbaro, one of Cornaro's circle and one of the era's great humanists. It was to Barbaro that Cornaro addressed the fourth chapter of *La Vita Sobria*.

Walking about in Cornaro-land during the lunch hour, one also encounters a certain aroma, the seductive smell of the greatest cuisine in the world. One of those smells is that of *brodo*, or homemade broth. To my mind, perhaps steeped too long in the past, it conjured Alvise Cornaro's eternal "dear" soup, *panatella*, or *panado*—that simple concoction of broth, small pieces of whole-grain bread, some vegetables, and perhaps an egg. If any one thing stands in light of all the modern science, it's the *panado*, a nutrient-heavy, calorie-light food that, simply put, *fits*. Its fits energetics theory. It fits insulin sensitivity theory. It fits commonsense theory. It even works metaphorically. In our ever-fraught energy-dense world, its message is:

eat light, live light, live long. In the grand, longevity-enhanced world of the twenty-first century and beyond, I could see it: Alvise's eternity soup, boiling away in the convivial kitchens of the world, and tasting pretty darn good.

Especially if you add a little grated Parmesan.

RECIPES FOR ETERNITY

Eternity Soup One: Cornaro's Broth

Have your cook slay and pluck a young capon; alternately, attend to your local market and buy one. Plunge the cleaned bird into a large pot of boiling water. Skim the resultant foam, then lower the heat and allow it to simmer for four hours. Remove the bird, cool and sieve the broth.

Have your baker make a small loaf of whole wheat bread; alternately, attend to your local market and purchase one. Pull it apart, allow it to dry a bit, then put small bits into the broth. Return to a vigorous boil for 5 minutes. Ladle the soup into a bowl. Add one tablespoon olive oil. Sprinkle with Parmesan. Eat, with one glass red or white wine.

Eternity Soup Two: The Frenchman's Broth

Having routed the army of King Francis I near the fraternal city of Pavia, it is rumored that a family of peasants in that gentle land gave succor to that King for two weeks, nourishing him with a garlic broth topped with a fatty toast and a poached egg. Deeming this soup *Zuppa di Pavia*, I have stripped it of its insipid Gallic fattiness and render preparations for such here.

Direct your cook to harvest and peel 15 cloves of fresh garlic. In a medium pan of water, boil these in a lively fashion for one hour. Direct your baker to slice, not too thickly, some fresh wheat bread. Scorch it in a stone *forno* or in another pan. Obtain a fresh egg, and crack its contents into the boiling broth and poach, one minute. Pour broth into a bowl, set the *tostato* on its surface, and then the poached egg. To flavor, add one tablespoon olive oil and some parsley, or, perhaps to *honor* that King, some . . . *chervil*. Eat, with a Venetian white.

Eternity Soup Three: The Sard's Zuppa

Being informed of the remarkable longevity of those rough men of Sardinia, I took to appraising their diet, and discovered that their broths approximate mine, although with the appearance of a strange fruit from the New World known as the tomato. I render it nearly palatable here.

Instruct your kitchen gardener to obtain seeds for said *pomodoro* from one of the skulking Spaniards who leer about near the port, and then plant, wait for 70 days, and harvest the fruit when it is red. Alternately, purchase a small weight of them from the Trader Joe. Peel, slice, and place, along with a clove of garlic, in a medium pan of water and boil for one hour. Now *mash* this mixture well and sieve.

Direct your baker to cook a large, wheat flatbread, or consult, again, the Trader. Pull shards of this orb apart, place in the broth with five stamens of saffron, and boil five minutes. Crack one egg per person and poach. Serve broth, softened flatbread, and egg, topped with one tablespoon olive oil and grated Sardinian cheese. Eat with one small glass Mirto, the rousing Sard liquor made from myrtle.

Eternity Soup Four: German Magic Suppe

The Germans, wanting their nose in everyone's soup, have finally invented one apropos of their mechanical sensibilities. It is called a boullion cube. I have to report that it suffices, in haste, for my divine broth.

Take one cube and, with a fire under the pan, *smash* it with a wooden spoon, slowly adding two cups of water. Stir. Instruct the cook to toast one piece of good *Italian* country bread. Poach one egg in the mechanical "broth." Assemble the soup. Flavor with one tablespoon olive oil and, perhaps, a sprinkle of paprika. Eat with a glass of Liebfraumilch.

Eternity Soup Five: à la Virtu

Considering that so many are randy for the alleged properties of greens, I made up a broth to accommodate their inclination. That it is French in origin is simply something one must tolerate.

Have your gardener pick a trug full of young dandelions,

and then clean them well. You may also obtain bitter greens in the form of endive or mustard at the market. Run a pot half full of water and set to a boil. Plunge in the greens, along with 5 garlic cloves. Boil for 40 minutes. Drain through a sieve, reserving the greens and the broth separately. Now poach several quail eggs, or one from a chicken. Toast a piece of rough wheat bread. Top the brodo with the toast and the egg, and season with one tablespoon olive oil and some of the reserved greens, well minced. Eat with a small glass of pastis.

NOTES

INTRODUCTION

xv   *Even the somnolent FDA*   See "Bioidentical Hormones: Sound Science or Bad Medicine," in *Statement of Steven Galson, MD, PhD, Director, Center for Drug Evaluation and Research, FDA,* April 19, 2007, online at http://www.fda.gov/ola/2007/hormone041907.html.

xv   *The Gerontological Society of America*   J. Maxwell, J. D. Melman, et al., "Anti-Aging Medicine: Can Consumers Be Better Protected?" *The Gerontologist* 44(2004):304–10.

xv   *A group of Big G researchers*   H. Warner et al., "Science Fact and the SENS Agenda: What Can We Reasonably Expect from the SENS Agenda?" *EMBO Reports* 6, no. 11 (Nov. 2005):1006–8.

xvi   *As a graduate student named Adam Spong*   Adam Spong, interview by Greg Critser, Springfield, IL, April 29, 2008.

xvi   *Stanford University professor Shripad Tuljapurkar*   See "Demographic and Economic Consequences of Extended Lives," presented at Antiaging Therapy: Biological Prospects and Potential Demographic Consequences session at the annual meeting of the American Association for the Advancement of Science, St. Louis, MO, February 17, 2006.

xvii   *The National Institute of Aging*   See R. A. Miller et al., "An Aging Interventions Testing Program: Study Design and Interim Report," *Aging Cell* 6, no. 4 (Aug. 2007):565–75. Epub, June 18, 2007.

xvii   *Another NIA study*   L. Fontana et al., "Long-Term Low-Calorie Low-Protein Vegan Diet and Endurance Exercise Are Associated with Low Cardiometabolic Risk," *Rejuvenation Research* 10, no. 2 (June 2007):225–34.

xvii   *When the noted RAND Corporation*   Dana P. Goldman et al., "Health Status and Medical Treatment of the Future Elderly: Final Report," RAND Health (Santa Monica, CA: RAND Corp., 2004).

xvii   *The cofounder of Pay Pal*   Read the announcement of the donation at http://www.mprize.org.

xvii   *The pharmaceuticals giant Glaxo*   Andrew Pollack, "Glaxo Says Compound in Wine May Fight Aging," *New York Times*, April 23, 2008, p. B1.

xvii   *The head of Rice University's*   A. D. de Grey, P. Alvarez, et al., "Medical Bioremediation: Prospects for the Application of Microbial Catabolic Diversity to Aging and Several Major Age-Related Diseases," *Ageing Research Reviews* 4, no. 3 (Aug. 2005):315–38. See also Aubrey de Grey, *Ending Aging* (St. Martin's Press, 2007); interviews with de Grey and Alvarez by Greg Critser.

xviii   *The NIH is funding*   Burcin Ekser et al., "Xenotransplantation on Solid Organs in the Pig-to-Primate Model," *Transplant Immunology*, October 26, 2008.

Book One: Calories

1   *"We must resist"*   Marcus Tullius Cicero, "On Old Age," *Letters and Treatises of Cicero and Pliny* (New York: P. F. Collier and Son, 1909), p. 58.

1   *The speaker, Josh Mitteldorf*   Josh Mitteldorf, "Evolution and Calorie Restriction," speech before the CR Society, April 7, 2006.

2   *He had recently published*   Longo, V. J. Mitteldorf, et al., "Programmed Aging and Altrusim," *Nature Review Genetics* 6, no. 11 (Nov. 2005):866–72.

2   *"Can someone let me use their computer"*   Remarks made during "Evolution and Calorie Restriction."

3   *"You can learn all kinds of things . . ."*   Edward Masoro, interviews by Greg Critser, April 7, 2007, and May 22, 2008.

3   *"Usually they are eating . . ."*   Steve Spindler, interviews by Greg Critser, April 7, 2007, and Feb. 6, 2008.

4   *"These are wonderful people . . ."*   Luigi Fontana, interviews by Greg Critser, April 8, 2006 and Aug. 11, 2007.

4   *"I don't know what was worse . . ."*   April Smith, interview by Greg Critser, Nov. 8, 2007.

5   *The rest, as Lisa Walford*   Lisa Walford, interview by Greg Critser, April 9, 2006.

7   *On several pages McCay had jotted down*   Clive McCay, "Current Reading List Concerning Problems of Aging," typed manuscript compiled June 1950, in Clive McCay papers, box 10, Cornell University.

7   *I first came across his treatise*   Alvise Cornaro, *The Art of Living Long* (Milwaukee: William F. Butler, 1903). This edition is the most widely available among a number published in the United States over two centuries.

8   *Born in Venice in 1484*   The best single source for detail of Cornaro's life is to be found in Marisa Milani, *Scritti Sulla Vita Sobria* (Venice: Corbo e Fiore Editore, 1988), pp. 36–60.

8   *His symptoms were telling*   Cornaro, *The Art of Living Long*, p. 44.

9   *But as Cornaro himself*   Ibid., p. 78.

10   *"In less than a year . . ."*   Ibid., p. 47.

10   *"I am healthy, cheerful and contented . . ."*   Ibid., p. 67.

12   *In 1793, the Reverend Parson Weems*   Weems published *La Vita Sobria* under the title *The Immortal Mentor or Man's Unerring Guide to a Healthy, Wealthy and Happy Life in Three Parts by Lewis Cornaro, Dr. Franklin and Dr. Scott* (Trenton: Daniel Fenton, 1810).

13   *Just outside Burlington*   The miracle of the Internet allows one to see the place where the modern science of caloric restriction was born. See http://www.munic.state.ct.us/BURLINGTON/hatchery/index.html.

14   *As a boy he inhaled the natural*   J. K. Loosli, *Journal of Nutrition* 103, no. 1 (Jan. 1993):3.

15   *Right away, as Bing later recounted*   L. C. Bing, "Old *Salvelinus fontinalis,*" *Perspectives in Biology and Medicine* 13, no. 4 (Summer, 1979), excerpted in Jeanette B. McCay, *Clive McCay: Nutrition Pioneer* (Charlotte Harbor, FL: Tabby House, 1994), p. 142.

15   *"Most of the fish in the group . . ."*   Ibid.

16   *"This force is continuously used up . . ."*   Ibid., p. 143.

16   *"They behave . . ."*   Ibid. See also C. M. McCay et al., "Factor H in the Nutrition of Trout," *Science* 67, no. 1731 (March 1928):249–250.

16   *He "was going to work on longevity . . ."*   McCay, *Clive McCay*, p. 143.

17   *About this, McCay complained and*   Clive McCay, speech to New York State Farmers Dinner, New York, NY, January 19, 1954, cited in ibid., p. 484.

17   *Writing in the September 1934*   C. M. McCay and Mary F. Crowell, "Prolonging the Life Span," *Scientific Monthly*, Nov. 1934, pp. 405–14, cited in ibid., p. 486. See also the original study, C. M. McCay, Mary F. Crowell, and L. A. Maynard, "The Effect of Retarded Growth Upon the Length of Life Span and Upon the Ultimate Body Size," *Journal of Nutrition* 10, no. 1, (1935).

18   *The man once known for his nature boy*   Bing, "Old *Salvelinus fontinalis*," cited in McCay, *Clive McCay*, p. 143.

19   *Life span was not fixed . . .*   Clive McCay Radio Transcripts, cited in McCay, *Clive McCay*, p. 486.

20   *The book was by a man*   The popular fusion of McCay's life-long interest in nutrition and life span can be found in Clive and Jeanette McCay, *You Can Make Cornell Bread* (Englewood, FL: Jeanette McCay, 1955). Two outstanding examples of the further evolution of McCay's thinking about life span can be found in Clive McCay, *Notes on the History of Nutrition Research* (Berne, Switzerland: Hans Huber Publishers, 1973), pp. 14–23; and Clive McCay, "Effect of Restricted Feeding Upn Aging and Chronic Diseases in Rats and Dogs," *American Journal of Public Health* 37 (May 1947):521–528.

22   *Among them were elaborations of the*   For a succinct, precise description of aging theories, see Edward J. Masoro, *Challenges of Biological Aging* (New York: Springer, 1999), pp. 33–72; and Robert E. Ricklefs and Caleb E. Finch, *Aging: A Natural History* (New York: Scientific American Library, 1995), pp. 1–47 and pp. 127–179.

23   *His name was Denham Harman*   K. Kitani and G. O. Ivy, " 'I Thought, Thought, Thought for Four Months in Vain and Suddenly the Idea Came'—An Interview with Denham and Helen Harman," *Biogerontology* 4(2003):401–412.

24   *"I became interested . . ."*   Ibid, p. 402.

25   *"I felt there had to be some common . . ."*   Ibid., p. 403.

25   *Then, around November 9 . . .*   Ibid., p. 404.

26   *"Free radicals cause . . ."*   Ibid. See also Denham Harman, "Free Radical Theory of Aging: An Update," *Annals of the New York Academy of Sciences* 1067(2006):10–21.

27   *Harman still does not*   Kitani and Ivy, "I Thought," p. 409.

27   *Leonard Hayflick, a driven young*   For an outstanding profile of Hayflick, see Stephen S. Hall, *Merchants of Immortality* (Boston: Houghton Mifflin, 2003), pp. 14–41, as well as Leonard Hayflick, *How and Why We Age* (New York: Ballantine Books, 1994), especially "Aging Under Glass," pp. 111–137.

28   *"I knew I had something then"*   Quoted in Suresh I. S. Rattan, " 'Just a Fellow Who Did His Job . . . ', an Interview with Leonard Hayflick," *Biogerontology* 1(2000):79–87, 81.

28   *As Hayflick recalled*   Ibid.

29   *"We wrongly assume . . ."*   Ibid., p. 86.

30   *"Humans' ability to tamper . . ."*   Ibid. For an elaboration of Hayflick's views on antiaging medicine, see Leonard Hayflick, "Aging: The Reality—Anti-Aging Is an Oxymoron," *Journals of Gerontology Series A: Biological Sciences and Medical Sciences* 59(2004):B573–B578.

30   *"I don't know why Len and that gang . . ."*   Interview with Dr. Stephen Spindler, University of Riverside, by Greg Critser, Feb. 6, 2008.

30   *"I found it to be an unworthy hypothesis . . ."*   Interview with Edward Masoro by Greg Critser, May 22, 2008.

30   *Edward Masoro evoked*   An outstanding profile of Masoro can be found in Ingfei Chen, "Hungry for Science," *Science of Aging Knowledge Environment*, 2003, no. 1, (Jan. 2003):nf1.

31   *His pet peeve*   Edward Masoro, comments to the Caloric Restriction Society, San Antonio, TX, Nov. 9, 2007.

31   *"The more I heard . . ."*   Interview with Ed Masoro by Greg Critser, May 22, 2008.

32   *To deduce exactly*   See, for example, the following: H. Bertrand, B. Yu, and E. Masoro, "The Effect of Rat Age on the Composition and Functional Activities of Skeletal Muscle Sarcoplasmic Reticulum

Membrane Preparations," *Mechanisms of Ageing and Development* 14, no. (Jan.–Feb. 1975):7–17; J. Stiles, A. Francendese and E. Masoro, "Influence of Age on Size and Number of Fat Cells in the Epididymal Depot," *American Journal of Physiology* 229, no. 6 (Dec. 1975):1561–8; E. Masoro, H. Bertrand and B. Yu, "Action of Food Restruction in Delaying the Aging Process," *Proceedings of the National Academy of Sciences (US)* 79, no. 13 (July 1982):4239–41.

33   *Instead, Masoro showed that*   Masoro et al., "Action of Food," p. 4240. For his and Austad's ideas about hormesis, aging, and CR, see Masoro and Austad, "The Evolution of the Anti-Aging Action of Dietary Restriction: A Hypothesis," *Journal of Gerontology Series A: Biological Sciences and Medical Sciences* 51, no. 6 (Nov. 1996):B387–91.

34   *The more he looked at aging, disease*   Edward Masoro, interview by Greg Critser, May 22, 2008. See also E. J. Masoro, "Aging: Current Concepts," *Aging (Milano)* 9, no. 6 (Dec. 1997):436–7.

35   *"It is totally arbitrary . . ."*   Interview with Ed Masoro by Greg Critser, May 22, 2008.

35   *"I always thought CR was . . ."*   Ibid.

36   *From an early age*   Lisa Walford, interview by Greg Critser, Santa Monica, CA, April 2, 2006.

36   *In an article in his high school's*   Roy Walford, *Maximum Life Span* (Norton: New York, 1983), p. xi.

36   *There was his work on*   G. Troup, G. Smith, and R. Walford, "Life Span, Chronologic Disease Patterns, and Age-Related Changes in Relative Spleen Weights for the Mongolian Gerbil (*Meriones unguiculatus*)," *Experimental Gerontology* 4, no. 3 (Sept. 1969):139–43; R. Liu, B. Leung and R. Walford, "Effect of Temperature-Transfer on Growth of Laboratory Populations of a South American Annual Fish *Cynolebias bellottii*," *Growth* 39, no. 3 (Sept. 1975):337–43; R. Walford and W. Hildemann, "Life Span and Lymphoma-Incidence of Mice Injected at Birth with Spleen Cells Across a Weak Histocompatibility Locus," *American Journal Pathology* 47, no. 5 (Nov. 1965):713–21.

36   *As a longtime colleague put it*   Interview with Stephen Spindler by Greg Critser, Feb. 6, 2008.

37   *"With a fairly high order of . . ."*   R. L. Walford, "The Exten-

sion of Maximum Lifespan," *Clinical Geriatric Medicine* 1, no. 1 (Feb. 1985): 29–35.

37 *He wrote a book*   Roy Walford, *The 120 Year Diet: How to Double Your Vital Years* (Avalon: New York, 2000). For the recipe to "Sherm's Mega muffins," see http://en.wiki.calorierestriction.org/index.php/Sherm's_Megamuffins.

37 *One of his best friends*   Interview with Caleb Finch by Greg Critser, Nov. 2, 2007. Also see Finch's appreciation upon Walford's death in "Dining with Roy," *Experimental Gerontology* 39(2004):893–4.

38 *You did not know what you were*   Roy Walford, "Biosphere 2 as a Voyage of Discovery: The Serendipity from Inside," *BioScience* 52, no. 3 (2002):259–63.

39 *"They started calling him . . ."*   Interview with Stephen Spindler, University of California, Riverside, Feb. 6, 2008.

39 *"It was very unpleasant when we . . ."*   Interview with Ed Masoro by Greg Critser, May 22, 2008.

39 *Studying the crew's blood tests*   Walford et al., "Calorie Restriction as Viewed from Biosphere 2," *Receptor* 5, no. 1 (Spring 1995):29–33; Walford et al., "The Calorically Restricted Low-Fat Nutrient-Dense Diet in Biosphere 2 Significantly Lowers Blood Glucose, Total Leukocyte Count, Cholesterol, and Blood Pressure in Humans," *Proceedings of the National Academy of Sciences (US)* 89, no. 23 (Dec. 1992):11533–7.

39 *Its concern was reflected*   Walford et al., "Atypical Parkinsonism and Motor Neuron Syndrome in a Biosphere 2 Participant: A Possible Complication of Chronic Hypoxia and Carbon Monoxide Toxicity?" *Movement Disorders* 19, no. 4 (April 2004):465–9.

40 *"I mean, I was attracted to Michael . . ."*   April Smith, "How Increased Involvement of Women Is Changing the Public Perception of CR, the CR Society, and Individual Experiences of CR," speech to the Caloric Restriction Society, San Antonio, TX, Nov. 5, 2007.

42 *"I became aware of aging . . ."*   Interview with David Fisher by Greg Critser, Nov. 7, 2007, and e-mail correspondence, March 7, 2008.

42 *"It eventually dawned on me . . ."*   Interview by Greg Critser, Nov. 6, 2007, and e-mail correspondance with "Kevin," March 6, 2008.

44 *I was there at the invitation of*   See http://www.calorierestriction

.org/files/documents/CR-Society_GC_Workshopt.pdf for an overview of the event.

45 *All of this had led McGlothin* Interview with Paul McGlothin by Greg Critser, Aug. 8, 2007.

45 *". . . the old image, you know . . ."* Interview with Paul McGlothin, Aug. 10, 2007.

45 *"At 59, he should have energy . . ."* Interview and e-mail correspondence with David Harrison, Sept. 2, 2007. See also my blog entry on the subject at www.Scientificblogging.com, under "zoology."

46 *"You've got to control . . ."* Paul McGlothin, "Introduction to Glucose Control," speech before the Caloric Restriction Society, Tarrytown, NY, Aug. 10, 2007. All further quotes by McGlothin were in speeches and comment sessions at this same conference. For an elaboration of his views, see Paul McGlothin and Meredith Averill, *The CR Way* (New York: Collins Living, 2008).

Book Two: Cash

53 *"Well, now, Doctor . . ."* Op. Cit. in Theodore B. Schwartz, "Henry Harrower and the Turbulent Beginnings of Endocrinology," *Annals of Internal Medicine* 131(1999):702–706.

53 *The first, entitled "The Six Stages of the Life Course"* George M. Martin and Caleb E. Finch, "An Overview of the Biology of Aging: A Human Perspective," in Guarente, Partridge, and Wallace, *Molecular Biology of Human Aging* (Cold Spring Harbor, NY: Cold Spring Harbor Press, 2008), p. 116.

54 *The second—a pair of original illustrations* Georges Debled, *Au-dela de cette limite votre ticket est toujours valuable* (Paris: Editions Albin Michel, 1992), p. 85.

54 *"grumpiness" as a medical term* See M. Haren et al., "Effect of 12 month oral testosterone on testosterone deficiency symptoms in symptomatic elderly males with low-normal gonadal status." *Age and Ageing* 2005 Mar;24(2):125–30. Epub 2004 Dec 13.

55 *"Rich guys playing with their hormones"* Fran and Neal Kaufman, interview by Greg Critser, Brentwood, CA, Jan. 20, 2008.

57 *Since the late nineteenth century* For an overview, see John

Hoberman, *Testosterone Dreams: Rejuvenation, Aphrodisia, Doping* (University of California Press, 2006); and Elizabeth Siegel Watkins, *The Estrogen Elixir: A History of Hormone Replacement Therapy in America* (Baltimore and London: JHU Press, 2007).

58   *"They were just too frightened of . . ."*   Robert Butler, interview by Greg Critser, New York, April 14, 2008.

59   . . . *"hard-ons"*   John Grasela, interview by Greg Critser, March 19, 2008. San Diego.

62   *"It made sense for a guy like me . . ."*   Ron Rothenberg, interview by Greg Critser, La Jolla, CA, Jan. 16, 2008.

63   . . . *the highly flaccid chart illustrating that drop*   Thomas Travison et al., "A Population-Level Decline in Serum Testosterone Levels in American Men," *Journal of Clinical Endocrinology and Metabolism* 92, no. 1 (2007):196–202.

68   *"The effects of six months of human growth hormone . . ."*   Daniel Rudman et al, "Effects of Human Growth Hormone on Body Composition in Elderly Men," *Hormone Research* 36, Suppl. 1 (1991):73–81.

69   *In the 1980s, few were more enterprising*   See their résumés at www.worldhealth.net.

70   *"It was very exciting . . ."*   Interview of Vincent Giampapa by Greg Critser, Jan. 8, 2008.

70   *"Also present at the [Cancun] conference . . ."*   Ronald Klatz, *Grow Young with HGH* (New York: Harper Perennial, 1997), p. xviii.

71   *"We believe the consequences of not acting . . ."*   Ibid., p. 28.

72   *"Has the guy ever treated . . ."*   Interview with Ron Rothenberg by Greg Critser Jan. 16, 2008.

73   *"It was kind of an endocrinologist's dream"*   Andrzej Bartke, interview by Greg Critser, Springfield, IL, April 28, 2008.

74   *"We knew we could fix that . . ."*   Ibid.

75   *The results of the Bartke trial*   Andrzej Bartke, "Can Growth Hormone (GH) Accelerate Aging? Evidence from Transgenic Mice," *Neuroendocrinology* 78(2003):210–16; see also Ingfei Chen, "In Aging, Being Small May Have Its Advantages," *New York Times*, August 14, 2004.

75   . . . *"this was a remarkable piece of work."*   Richard Miller, interview by Greg Critser, Feb. 7, 2008.

75   *"They wrote this incredibly detailed . . ."*   Interview with Bartke, April 29, 2008.

75   *"Almost every one of these supposedly solid . . ."*   Ibid.

76   *"And remember"*   Ibid.

78   . . . *"We suspect that research efforts . . ."*   Michael S. Bonkowski, Andrzej Bartke, et al., "Targeted Disruption of Growth Hormone Receptor Interferes with the Beneficial Actions of Calorie Restriction," *PNAS* 103, no. 20(2006):7904.

79   *"That anyone can sit there and actually use . . ."*   Jay Olshansky, comments to Greg Critser at the opening session of the Gerontological Society of America annual convention, San Francisco, Nov. 16, 2007; e-mail correspondence with Critser, Dec. 19, 2007.

80   *Perhaps the most negative and influential*   Hau Liu, "Systematic Review: The Safety and Efficacy of Growth Hormone in the Healthy Elderly," *Annals of Internal Medicine* 146, no. 2 (Jan. 2007):104–15.

81   *A much-vaunted study of 384*   Nir Barzilai et al., "Functionally Significant Insulin-like Growth Factor I Receptor Mutations in Centenarians," *Proceedings of the National Academy of Science (US)* 105, no. 9 (March 2008):3438–42.

82   *"There is a huge and legitimate . . ."* Steven Austad, interview by Greg Critser, San Antonio, TX, Feb. 12, 2008.

82   *"I'm back on it . . ."*   Ron Rothenberg, interview by Greg Critser, June 11, 2008.

85   *I first met Hertoghe*   Thierry Hertoghe, "Diagnosing Hormone Deficiencies—A Live Consultation," speech and demonstration before the annual convention of American Academy of Anti-Aging Medicine, Las Vegas, Dec. 13, 2007.

87   *"We used to call them endocrine pearls"*   Fran Kaufman, e-mail to Greg Critser, Jan 11, 2008.

88   *"In Brussels there has been . . ."*   Interview with Thiery Hertoghe by Greg Critser, Jan. 22, 2008.

90   *"She had the drawling voice . . ."*   Eugene Hertoghe, speech before the International Surgical Congress, April 13, 1914.

90   *"I actually never meant . . ."*   Herthoghe interview, Jan. 22, 2008.

91   … "*le silhouette de l'homme …*"   Debled, *Au-dela de cette limite votre ticket est toujours valuable* (Paris: Editions Albin Michel, 1992), p. 85.

91   "*He sat me down …*"   Hertoghe interview, Jan. 22, 2008.

92   "*Hertoghe had maybe fifteen people …*"   Grasela interview, March 19, 2008.

92   *One thing was utterly telling*   Thierry Hertoghe, "Practical Applications in Treating Adult Hormone and Nutritional Deficiencies," at Bio-Identical Hormone Replacement Seminar, Las Vegas, Feb 28, 2008.

93   "*There's a firm face …*"   Ibid.

95   … "*you will never look at yourself the same way. …*"   Thierry Hertoghe, interview by Greg Critser, Feb 28, 2008.

95   "*Oh, my God, …*"   Ibid.

96   "*Generally speaking, I find that Wright …*"   James Duke e-mail to Greg Critser, March 12, 2008.

97   "*We need someone to sponsor …*"   Comments by Jonathan Wright to Bio-Identical Hormone Replacement Seminar, Las Vegas, Feb 28, 2008.

98   "*They didn't want us to even see …*"   Jonathan Wright, interview by Greg Critser, April 24, 2008.

99   *The findings were clear*   For a full description of this history, see Watkins, *The Estrogen Elixir.*

99   "*But that was a wrong assumption*"   Wright interview.

101   *without the increased risk of cancer and heart attack*   The controversy continues to rage. In 2009, researchers published a follow-up to the original Women's Health Initiative (WHI) report. The new study looked at women who discontinued HRT after the 2002 WHI report; it showed an apparent causal relationship between declines in breast cancer and discontinuation of HRT. See: Rowan T. Chlebowski et al, "Breast Cancer after Use of Estrogen plus Progestin in Postmenopausal Women," *New England Journal of Medicine* 360, 6 Feb. 5, 2009, pp. 573–85. Antiaging critics note that the new study has the same flaws as previous warnings: Almost two-thirds of the subjects were women commencing HRT after age sixty, rather than before or upon menopause; that they were using commercial PremPro, made from mare's urine, rather than bioidenticals; and that the dosages used by study participants were unnecessarily high, compared to lower dose bioindenticals.

101   *Harrower, a big, handsome man*   For a delightful description, see Theodore B. Schwartz, "Henry Harrower and the Turbulent Beginnings of Endocrinology," *Annals of Internal Medicine* 131(1999):702–706.

102   *"Surely nothing will discredit . . ."*   Op. Cit., ibid.

103   *. . . "so what is there to say on pluriglandular . . ."*   Ibid.

103   *. . . "This presidential diatribe . . ."*   Ibid. For the full work, see Harvey Cushing, "Disorders of the Pituitary Gland," *Journal of the American Medical Association* 76, no. 25 (June 1921):1722–26.

103   *"We did this for three to four rounds . . ."*   Interview with Wright.

104   *Pointing to slides*   David Steenblock, "Bone Marrow Stem Cell Therapy: A Major Breakthrough for Chronic Diseases and Anti-aging," speech to SENS3 Conference, Sept. 8, 2008. To view his entire presentation, see http://richardjschueler.com/gallery2/main.php?g2_itemId=57090.

105   *Nevertheless, he has been investigated*   "Petition for Order to Show Cause Re Appearance and Testimony Pursuant to Investigational Subpoena," Superior Court of California for the Country of Orange, No. 07CC00725, April 5, 2007.

105   *. . . Umbilical Cord Stem Cell Therapy*   David Steenblock and Anthony G. Payne, *Umbilical Cord Stem Cell Therapy* (Laguna Beach: Basic Health, 2006).

106   *Patton's claim was that TA-65*   Noel Patton, "Neutraceutical Telomerase Stimulation with TA-65," speech to SENS 3, September 6, 2007. See the video at http://richardjschueler.com/gallery2/main.php?g2_itemId=57036.

106   *"We are constantly checking . . ."*   Noel Patton, interview by Greg Critser, Las Vegas, December 14, 2007.

106   *"Who is he?"*   Interview with Rita Effros by Greg Critser, Sept. 19, 2007.

106   *. . . "looking for a way to affect human aging . . ."*   Vincent Giampapa, interview by Greg Critser, Jan. 8, 2008.

107   *"These tests are, at best . . ."*   Arlan Richardson, e-mail communication to Greg Critser, Aug. 29, 2008.

107   *"We are seeing mammals that . . ."*   Arlan Richardson, interview by Greg Critser, San Antonio, TX, Feb. 12, 2008. For a critical examination of the 8-oxo-2 oxidative stress test, see Michelle L. Hamilton, Arlan

Richardson, et al., "A Reliable Assessment of 8-oxo-2-deoxyguanosine Levels," *Nucleic Acids Research* 29, no. 10(2001):2117–26.

108   *In his classic work*   Masoro, *Challenges of Biological Aging*, pp. 89–94.

109   *"I was seeing . . ."*   John Shieh, interview by Greg Critser, South Pasadena, CA, Dec. 24, 2008.

112   *He replies, "It could be that the Testosterone/Estrogen . . ."*   Ron Rothenberg, in e-mail consultation with Greg Critser, July 23, 2008.

BOOK THREE: ENGINEERING

115   *"The great choice for this generation . . ."*   David Gobel, interview by Greg Critser, Cambridge, England, Sept. 7, 2007.

117   . . . *"there was never a time . . ."*   Joseph Vacanti, interview by Greg Critser, Boston, MA, April 16, 2008.

118   *"I watched in agony . . ."*   Joseph Vacanti, "Tissue Engineering and Regenerative Medicine, The Jayne Lecture," Nov. 15, 2003, in *Proceedings of the American Philosophical Society* 151, no. 4 (Dec. 2003):397–8. For an example of his work, see also Joseph P. Vacanti, "Tissue and Organ Engineering: Can We Build Intestine and Vital Organs?" *Journal of Gastrointestinal Surgery* 7, no. 7 (Nov. 2003):831–5, and E. R. Ochoa and J. P. Vacanti, "An Overview of the Pathology and Approaches to Tissue Engineering," *Annals of the New York Academy of Sciences* 979(2002):10–26; discussion 35-8.

118   *"The research question was in a way . . ."*   Vacanti interview, April 16, 2008.

119   *"The solution was staring me right in the face! . . ."*   Vacanti, "Tissue Engineering," p. 399.

119   *"All of this we did, really, as engineers"*   Vacanti interview, April 16, 2008.

120   *"Most organs are just a bunch of tubes"*   Gabor Forgacs, interview by Greg Critser, April 2, 2008.

120   . . . *"a theoretical background"*   John Critser, interview by Greg Critser, April 2, 2008.

121   *"I'm not big on scaffolds. . . "*   Forgacs interview, April 2, 2008.

122   *"Yes, I have a dream . . ."*   Ibid.

122   *In the summers of 2006 and 2007*   Shinya Yamanaka, "Induction of Pluripotent Stem Cells from Mouse Embryonic and Adult Fibroblast Cultures by Defined Factors," *Cell* 126, no. 4 (Aug. 2006): 663–76; then see A. Meisner et al., "Direct Reprogramming of Genetically Unmodified Fibroblasts into Pluripotent Stem Cells," *Nature Biotechnology* 25, no. 10 (Oct. 2007):1177–81.

122   *In 2008, researchers at Harvard*   Douglas Melton, "In Vivo Reprogramming of Adult Pancreatic Exocrine Cells to Beta-Cells," *Nature* 455, no. 7213 (Oct. 2008):627–32.

123   *"For most of our lives, endogenous . . ."*   Doris Taylor, interview by Greg Critser, March 11, 2008.

123   *In a kind of midwestern Frankenstein moment*   Doris A. Taylor et al., "Perfusion-Decellularized Matrix: Using Nature's Platform to Engineer a Bioartificial Heart, *Nature Medicine* 14, no. 2 (Feb. 2008):213–21.

124   *"Can stem cells be placed anywhere in the body . . ."*   Taylor interview.

126   *"China will overtake us . . ."*   John Critser interview.

126   *"the real barrier . . ."*   Chris Mason, "Regenerative Medicine 2.0," speech before SENS 3, Cambridge, England, Sept. 8, 2007. See a video of his presentation at http://richardjschueler.com/gallery2/main.php?g2_itemId=57322.

128   *"The prospects for continued increase . . ."*   Caleb E. Finch, "Variations in Senescence and Longevity Include the Possibility of Negligible Senescence," *Journal of Gerontology Biological Sciences* 53A, no. 4 (1998):B238, emphasis mine.

129   *"The ones who [eventually] do [breed] . . ."*   Quoted in Josie Glausiusz, "Michael Rose Beating Death," *Discover* 22, no. 05 (May 2001).

129   *"There are all kinds of people who are opposed . . ."*   Ibid.

130   *In 1991, Cynthia Kenyon*   C. Kenyon et al. "A. *C. elegans* Mutant That Lives Twice as Long as Wile Type." Nature. 1993 Dec. 2:366(6454):461–4.

130   *"These animals are magical"*   Cythnia Kenyon, "Laboratory Overview," http://kenyonlab.ucsf.edu/html/lab_overview.html.

130   *"The field of aging research . . ."*   Ibid., emphasis mine.

132   *"the long-lived mammals that Miller . . ."*   Aubrey D. N. J. de Grey, "Resistance to Debate on How to Postpone Ageing Is Delaying Progress and Costing Lives," *EMBO Reports* 6, Suppl. 1 (July 2005): S49–S53.

132   *"You don't really want to talk about my mice . . ."*   Richard Miller, interview by Greg Critser, Feb. 7, 2008.

133   *Over a period of about ten years*   For a comprehensive overview of his work and approach, see Aubrey de Grey and Michael Rae, *Ending Aging: The Rejuvenation Breakthroughs That Could Reverse Human Aging in Our Lifetime* (New York: St. Martin's Press, 2007).

133   *"It kills fucking 100,000 people . . ."*   Aubrey de Grey, "TED 2006 Conference Presentation," May 1, 2006.

134   *. . . "farrago" and a "fantasy"*   H. Warner et al., "Science Fact and the SENS agenda. What Can We Reasonably Expect from Ageing Research?" *EMBO Reports* 6, no. 11 (Nov. 2005):1006–8.

134   *Personal attacks followed*   Sherwin Nuland, "Do You Want to Live Forever?" *Technology Review*, February 2005. See my recap of this in Greg Critser, "The Man Who Will Help You Live for 1000 Years," *Times of London*, Sept. 7, 2007.

135   *"One reason," he told me*   Aubrey de Grey, interview one by Greg Critser, July 5, 2007.

138   *"The goal is to wipe out the damage . . ."*   Ibid.

139   *"One potential side effect . . ."*   de Grey and Rae, *Ending Aging*, p. 308.

139   *"I've never been able to not see . . ."*   De Grey interview one.

139   *"One fateful day . . ."*   Aaron Turner, "MMM and the SMVS Project," Feb. 19, 2008, found at http://www.manmademinions.com/mmm_smvs_backgrounder.pdf.

140   *What amazed, and later infuriated*   De Grey interview one.

142   *"These are lifelong injuries . . ."*   Rutledge Ellis-Behnke, "Using Nanotechnology to Repair the Body," presentation to SENS 3, Cambridge, England, Sept. 7, 2007. See the video at http://richardjschueler.com/gallery2/main.php?g2_itemId=57031.

143   *"With the pyramids . . ."*   Ibid.

145   *"At first we were just thinking . . ."*   Pierre Moreau, "Can We Influence the 'Normal' Aging of Arteries?," presentation to SENS 3, Cambridge, England, Sept. 6, 2007. See the video at http://richardjschueler.com/gallery2/main.php?g2_itemId=57021.

146   *For the past ten years or so*   Kim Janda, "Immunopharmacotherapy Against Weight Gain," presentation to SENS 3, Cambridge, England,

Sept. 7, 2007. See the video at http://richardjschueler.com/gallery2/main
.php?g2_itemId=57120.

149   *The field of bioremediation, Alvarez explained*   Pedro Alvarez, interview by Greg Critser, Houston, Nov. 14, 2007.

150   *"I mean, right now, in my lab . . ."*   Ibid. See also Aubrey de Grey
and P. Alvarez, "Medical Bioremediation: Prospects for the Application of
Microbial Catabolic Diversity to Aging and Several Major Age-Related diseases," *Ageing Research Reviews* 4, no. 3 (Aug. 2005):315–38.

151   *One of his associates had just*   A. M. Cuervo and C. Zhang, "Res-
toration of Chaperone-Mediated Autophagy in Aging Liver Improves Cellular Maintenance and Hepatic Function," *Nature Medicine* 14, no. 9 (Sept.
2008):959–65.

152   *"Human evolution will soon be . . ."*   Aubrey de Grey e-mail to
Greg Critser, August 23, 2008.

152   *"People really go into a sort of pro-aging . . ."*   De Grey interview
one.

153   *"Stage three"*   Aubrey de Grey, interview two by Greg Critser,
August 4, 2008.

BOOK FOUR: THE NEW LONGEVITY BESTIARY

158   *He sees aging as the body's*   Steven Austad, *Why We Age: What
Science is Discovering About the Body's Journey Through Life* (New York: John
Wiley, 1997). For a detail of his longevity quotient idea, see pp. 83–86.

159   *"I remember somebody asking Richard . . ."*   Steven Austad, interview by Greg Critser, San Antonio, TX, Feb. 11, 2008.

160   *"They live somewhere . . ."*   Ibid.

161   *"The problem with birds is that . . ."*   Ibid.

162   . . . *"Look at him. He's twenty-eight . . ."*   Rochelle Buffenstein,
interview by Greg Critser, San Antonio, TX, Feb. 12, 2008.

163   *"It's the kind of oxidation . . ."*   Ibid.

163   *"As you know, rats and mice . . ."*   Ibid.

164   *"Extended longevity is also correlated . . ."*   R. Buffenstein, "Negligible Senescence in the Longest-Living Rodent, the Naked Mole Rat,"
Journal of Comparative Physiology 178, no 4 (May 2008):439–45.

165   *"They lived just as long!"*   Rochelle Buffenstein, interview by Greg Critser, San Antonio, TX, Feb. 12, 2008.

167   *"Elephants contain about . . ."*   Austad interview.

167   *"This, to us, has been a huge shock"*   Randy Strong, interview by Greg Critser, San Antonio, TX, Feb. 13, 2008. For an overview of the testing program, and findings, see R. A. Miller et al., "An Aging Interventions Testing Program: Study Design and Interim Report," *Aging Cell* 6, no. 4 (Aug. 2007):565–75.

168   *"If you want to be flippant about it . . ."*   Suzette Tardif, interview by Greg Critser, San Antonio, TX, Feb. 12, 2008.

169   *Other scholars have shown*   B. K. Berkovitz and J. Pacy, "Age Changes in the Cells of the Intra-articular Disc of the Temporomandibular Joints of Rats and Marmosets," *Archives of Oral Biology* 45, no. 11 (Nov. 2000):987–95.

169   *Tardif already has linked*   S. Tardif et al., "Energy Restriction Initiated at Different Gestational Ages Has Varying Effects on Maternal Weight Gain and Pregnancy Outcome in Common Marmoset Monkeys (*Callithrix jacchus*)," *British Journal of Nutrition* 92, no. 5 (Nov. 2004):841–9.

169   *"Yes, low IGF-1 in transgenic mice . . ."*   Tardif interview. For an outstanding, state-of-the-art discussion of frailty, aging, and IGF-1, see Arnold J. Kahn, "Meeting Report: Central and Peripheral Mechanisms of Aging and Frailty," *Journal of Gerontology Biological Sciences* 62A, no. 12 (2007):1357–60.

BOOK FIVE: ONE MILLION CENTENARIANS

173   *"[Regarding the nutrition of older people] . . ."*   Clive McCay, "Current Reading List Concerning Problems of Aging," typed manuscript compiled June 1950, in Clive McCay papers, box 10, Cornell University.

173   *If you look at the work of people*   Shripad Tuljapurkar et al., "Demographic and Economic Consequences of Extended Lives," paper presented to the session on anti-aging therpy: biological prospects and potential demographic consequences, AAAS Annual Meeting, St. Louis, Feb. 17, 2006.

174   *"People are going to do things . . ."*   Ibid.

•

174   *Take them and put them in a room*   Goldman et al., "Health Status and Medical Treatment of the Future Elderly."

176   *He had a huge and often lively*   Robert N. Butler, *The Longevity Revolution: The Benefits and Challenges of Living a Long Life* (New York: Public Affairs, 2008).

177   *"It is amazing"*   Robert Butler, interview by Greg Critser, New York, April 14, 2008.

178   *"It's the science"*   Ibid.

179   *All of this, he said*   Ibid.

179   *The Russian Nobelist Ilya Metchnikoff*   Quoted in Butler, *The Longevity Revolution*, p. 254.

182   *On September 2, 2006*   For an outstanding PowerPoint on Johnson's life, see L. Stephen Coles, "The Secrets of the Oldest Old," *Edmonton Aging Symposium,* March 31, 2007, at http://www.edmontonagingsympo sium.com/files/eas/presentations/31-Stephen_Coles.ppt.

184   *"When I found out"*   Brian Johnson, interview by Greg Critser, Richmond, CA, Sept. 5, 2006.

188   *"everyone on earth who lives beyond . . ."*   L. Stephen Coles, interview by Greg Critser, Los Angeles, Feb. 7, 2008.

189   *"Look at that"*   L. Stephen Coles interview.

190   *"See," he said, clicking*   Harry Vinters, interview by Greg Critser, Los Angeles, April 23, 2008.

190   *"There was always someone staying . . ."*   Brian Johnson interview.

192   *Cornaro had one solution*   Cornaro, *La Vita Sobria*, p. 47.

193   *In his vaulted music room*   See recent photographs of the restored Cornaro residence in Vittorio Sgarbi, *L'Odeo Cornaro* (Torino: Umberto Allemandi, 2003).

SELECT BIBLIOGRAPHY

Austad, Steven. *Why We Age: What Science Is Discovering About the Body's Journey Through Life*. New York: John Wiley & Sons, 1997.

Butler, Robert. *The Longevity Revolution: The Benefits and Challenges of Living a Long Life*. New York: Public Affairs, 2008.

Butler, Robert. *Why Survive? Being Old in America*. New York: Harper&Row, 1975.

Cicero, Marcus. *On Old Age and On Friendship*. Translated, and with an introduction, by Frank Copley. Ann Arbor: Ann Arbor Paperback, 1978.

Cornaro, Alvise. *The Art of Living Long* [La Vita Sobria]. Milwaukee: William F. Butler, 1903.

Cosgrove, Denis. *The Palladian Landscape*. University Park, PA: Pennsylvania State University Press, 1993.

Debled, Georges. *Au-dela de cette limite votre ticket est toujours valable*. Paris: Albin Michel, 1992.

de Grey, Aubrey, and Michael Rae. *Ending Aging: The Rejuvenation Biotechnologies That Could Reverse Human Aging in Our Lifetime*. New York: St. Martin's, 2007.

Finch, Caleb E. *The Biology of Human Longevity: Inflammation, Nutrition, and Aging in the Evolution of Lifespans*. New York: Elsevier, 2007.

Finch, Caleb E., and Robert E. Ricklefs. *Aging: A Natural History*. New York: Scientific American Library, 1995.

Gale, Ernest F. *The Chemical Activities of Bacteria.* London: University Tutorial Press, 1948.

Goldman, Dana P. *Health Status and Medical Treatment of the Future Elderly: Final Report.* Santa Monica, CA: RAND Health, RAND Corp., August 2004.

Gruman, Gerald J. *A History of Ideas About the Prolongation of Life.* New York: Springer, 2003.

Guarente, Lenny, *Ageless Quest: One Scientist's Search for the Genes That Prolong Youth.* New York: Cold Spring Harbor Laboratory Press, 2003.

Guarente, Leonard P., Linda P. Partridge, and Douglas C. Wallace, eds. *The Molecular Biology of Aging.* New York: Cold Spring Harbor Laboratory Press, 2008.

Hall, Stephen S. *Merchants of Immortality: Chasing the Dream of Human Life Extension.* Boston: Mariner Books, 2003.

Hayflick, Leonard. *How and Why We Age.* New York: Ballantine Books, 1994.

Hertoghe, Thierry. *The Hormone Solution: Stay Younger Longer.* New York: Three Rivers Press, 2002.

Huxley, Aldous. *After Many a Summer Dies the Swan.* New York: Avon, 1939.

Klatz, Ronald. *Grow Young with HGH.* New York: HarperPerennial, 1997.

Lee, John R. *Natural Progesterone: The Multiple Roles of a Remarkable Hormone.* Sebastopol, CA: BLL Publishing, 1993.

Logan, Oliver. *Culture and Society in Venice 1470–1790.* New York: Scribner's, 1972.

Masoro, Edward J. *Challenges of Biological Aging.* New York: Springer Publishing, 1999.

McCay, Clive. *Notes on the History of Nutrition Research.* Berne: Hans Huber, 1973.

McCay, Jeanette B. *Clive McCay: Nutrition Pioneer: Biographical memoires by his wife.* Charlotte Harbor, FL: Tabby House, 1994.

Milani, Marisa. *Scritti Sulla Vita Sobria: Elogia e Lettere*. Venice: Corbo e Fiori Editori, 1988.

Olshansky, S. Jay, and Bruce A. Carnes. *The Quest for Immortality: Science at the Frontiers of Aging*. New York: W. W. Norton, 2003.

Rothenberg, Ron, Kathleen Becker, and Kris Hart. *Forever Ageless: Advanced Edition*. La Jolla: California Healthspan Institute, 2007.

Sgarbi, Vittorio, and Lorenzo Capellini. *L'Odeo Cornaro*. Torino: Umberto Allemandi & Co., 2003.

Somers, Suzanne. *The Sexy Years*. New York: Crown, 2004.

Tafuri, Manfredo. *Venice and the Renaissance*. Translated by Jessica Levine. Cambridge, MA: MIT Press, 1995.

Walford, Roy. *The 120 Year Diet: How to Double Your Vital Years*. New York: Thunder's Mouth Press, 2000.

Walford, Roy. *Maximum Life Span*. New York: W. W. Norton, 1983.

Watkins, Elizabeth Siegel. *The Estrogen Elixir: A History of Hormone Replacement Therapy in America*. Baltimore: Johns Hopkins University Press, 2007.

Wright, Jonathan V., and John Morgenthaler. *Natural Hormone Replacement*. Petaluma, CA: Smart Publications, 1997.

Zerbi, Gabriele. *Gerontocomia: On the Care of the Aged* in Venice, *1489* in *Gabriele Zerbi, Gerontocomia: On the Care of the Aged* and *Maximianus, Elegies on Old Age and Love*. Translated from the Latin by L. R. Lind. Philadelphia: American Philosophical Society, 1988.

ACKNOWLEDGMENTS

It is said that books are acts of passion, logic, reason, and art. They are also works of faith—by those who support, tutor, guide, mentor, and succor the writer. This book greatly benefited from the abiding faith of three individuals. The first is Caleb "Tuck" Finch, professor of gerontology and biological science at the University of Southern California. There, I have been lucky enough to attend his classes, lectures, and discussions. Equally important outside of USC were the many conversations, e-mails, lunches, and dinners during which Tuck freely gave me his counsel and advice on a mind-boggling array of scientific issues regarding the biology of aging. I can't thank him enough. The second Virgil in this journey is my editor at Harmony Books, the ever-calm John Glusman, who deftly pointed out the most important signposts, and who gave me the resources and time to complete the trip. In the tumult of modern publishing, he's a port in the storm. The third believer was Richard Abate, my agent, who pushed me to stretch and always had time for my questions, both practical and, sometimes more important, transcendental. He also understands pizza.

A number of scholars were good enough to take time from their demanding schedules to educate me on their specialties. Chief among them was the biologist Steven Austad at the Barshop Institute for Longevity and Aging Studies at the University of Texas Health Science Center in San Antonio. If it weren't for him, I wouldn't know a naked mole rat from a cotton top tamarin. His fellows at Barshop also came through—James Nelson, Arlan Richardson, Suzette Tardif, Randy Strong, and Rochelle Buffenstein among them. At UCLA, I am indebted to Professor Rita Effros, the trailblazing pioneer in studies of immune system aging; to the cellular pathology expertise of Professor Harry Vinters; to the wisdom of medical school dean Gerald Levey and vice chancellor Barbara Levey; to the learned counsel in *Cornariana* of Massimo Ciavolella; to the encouragement and example of former chancellor Albert Carnesale; and, as always, to the librarians at the Louise Darling Biomedical Library. Fran Kaufman and Robert Binstock both provided encouragement and insight. For my understanding of laboratory animals I am indebted to Joyce Peterson and the Jackson Laboratories, and to the American Association of Laboratory Animal Scientists. Professor Andrzej Bartke, at the University of Illinois, tutored me in the fundamentals of human growth hormone and IGF-1. I still can't afford the stuff.

Several editors also helped. Among them: Sue Horton and Nick Goldberg at the *LA Times*, who let me write about Alzheimer's, aging, and Alvise Cornaro; Hank Campbell, who edited my zoology blog at Scientificblotting.com; and Gill Hemburrow at the *Times of London*, who edited my profile of Aubrey de Grey. Roger Hodge at *Harper's* let me write about lab animals. Michael Balter, a veteran writer at *Science*, acted as a valuable editorial sounding board and occasional gusta-tory companion, as did Steve Oney. Anne Berry at Harmony Books contributed mightily to the visual composition of the book.

Among those who have taken up the popular battle against aging, few were more helpful that Aubrey de Grey, PhD, of Cambridge, England. Members of the Caloric Restriction Society—most courageously Michael Rae and April Smith—granted me access to their meetings and their private lives as well. Thank you. I still think you need more pizza. The physician and antiaging advocate Ron Rothenberg agreed to treat me (no discount!) and to consent to a patient-journalist's critical scowl. Doctors Thierry Hertoghe and Jonathan Wright also tolerated my scrutiny. L. Stephen Coles, the founder of the Supercentenarian Research Project, provided a number of invaluable case studies, including much on the topic of George Johnson, who died during the research of this book at age 112. Lisa Walford was an invaluable source of insight about both caloric restriction and about her father, the late Roy Walford. My mother, Betty Critser, and my stepfather, Jerry Newman, were key wellsprings of this book. They keep looking better—and it's not caloric restriction!

One more thing: writers hog psychic space, and those who yield that space are the true, albeit hidden, benefactors of any work. My wife, Antoinette Mongelli, is this work's spiritual benefactor, and to her there is no acknowledgment big or last-ing enough. Thanks. I love you.

CREDITS

INDEX

Page numbers in *italics* refer to illustration captions.

About the Author

GREG CRITSER is a longtime journalist and observer of the medical industry. He is the author of *Fat Land: How Americans Became the Fattest People in the World* and *Generation Rx: How Prescription Drugs Are Altering American Lives, Minds, and Bodies.* His work has appeared in dozens of magazines and newspapers, and he is a well-known commentator on medicine, health, and food, with regular appearances on public radio and TV. He lives in Pasadena with his wife, Antoinette Mongelli.